SCIENCE, POLITICS AND MORALITY

THEORY AND DECISION LIBRARY

General Editors: W. Leinfellner (*Vienna*) and G. Eberlein (*Munich*)

Series A: Philosophy and Methodology of the Social Sciences

Series B: Mathematical and Statistical Methods

Series C: Game Theory, Mathematical Programming and Operations Research

Series D: System Theory, Knowledge Engineering and Problem Solving

SERIES A: PHILOSOPHY AND METHODOLOGY OF THE SOCIAL SCIENCES

VOLUME 17

Scope: This series deals with the foundations, the general methodology and the criteria, goals and purpose of the social sciences. The emphasis in the new Series A will be on well-argued, thoroughly analytical rather than advanced mathematical treatments. In this context, particular attention will be paid to game and decision theory and general philosophical topics from mathematics, psychology and economics, such as game theory, voting and welfare theory, with applications to political science, sociology, law and ethics.

The titles published in this series are listed at the end of this volume.

SCIENCE, POLITICS AND MORALITY

Scientific Uncertainty and Decision Making

edited by

RENÉ VON SCHOMBERG

University of Brabant, Tilburg, The Netherlands

Springer-Science+Business Media, B.V.

Library of Congress Cataloging-in-Publication Data

Science, politics, and morality : scientific uncertainty and decision making / edited by René von Schomberg.
p. cm. -- (Theory and decision library. Series A., Philosophy and methodology of the social sciences ; v. 17)
Includes index.

1. Science--Social aspects. 2. Technology--Social aspects. 3. Science and state. 4. Technology and state. 5. Decision-making. I. Schomberg, René von. II. Series.
Q175.5.S367 1993
303.48'3--dc20 92-31872

ISBN 978-90-481-4211-8 ISBN 978-94-015-8143-1 (eBook)
DOI 10.1007/978-94-015-8143-1

Printed on acid-free paper

ACKNOWLEDGMENTS

This book is a result of the workshop 'Science, Politics and Morality' held at the University of Twente on 5-8 June 1991. I prepared this workshop together with my cherished companion for life, Patricia Morales. I want to express my appreciation for the fact that the Faculty of Philosophy and Social Sciences of the University, and in particular Prof.dr. P.J.M. Schellens, to provide me the possibility and optimal conditions to organize the workshop. I am grateful for a fund of the foundation 'Stichting Fonds Landbouw Export Bureau 1916/1918' of the Agricultural University of Wageningen to contribute in the costs to produce the book. I also want to express my gratitude to Anneke Aarten who kindly helped with the lay-out of the publication. I am indebted to Prof.dr. A.M.T Meyer who extensively helped with the english idiom of my contribution to the book, but more than that, many fruitful ideas were developed in discussions with him on many occasions.

The editor, June 1992

CONTENTS

INTRODUCTION

René von Schomberg

The Value Spheres Science, Politics and Morality

In modernity we have experienced the differentiation of the *value spheres* (Max Weber) Science, Politics, Legality and Morality. In traditional social theory these value spheres are understood to be autonomous and mutually independent. They are claimed to be autonomous in the sense that the specific claims made within each value sphere, e. g. truth claims within science or judicial rightness claims within the legal system, are selected and developed in discourses of expert cultures specialised in the single and authoritative production of one type of claim. They are claimed to be independent in the sense that the development and progress within one value sphere can be achieved without appealing to the results of the other special discourses. The assumptions of modernity about the autonomy and independence of our value spheres have been heavily debated in the social sciences and the humanities. These debates resulted in the emergence of new theoretical approaches, like constructivism in the sociology of science, and in a revival of pre-modern authors in philosophy. One can wonder whether the fundamental problems of the differentiation of the value spheres are still taken seriously by these fashionable trends in science. In the papers collected for this book, we are not so much concerned with the challenge of discussing new approaches in science. Here we concentrate on how the social challenge to solve the problems of technological risks and the ecological crises effects the institutional arrangements of the value spheres science, politics and morality.

In part I, the papers of Von Schomberg, Kollek and Shrader-Frechette deal with the nature of the controversies about risks and its implication for the policy making process. Von Schomberg shows, by means of an analysis of argumentation, that in the case of epistemic discussions in science, we cannot reasonably expect a consensus among disputing experts from different scientific disciplines, since their arguments can not substantiate the truth of claims but, rather, do state their incompatible plausibility. The problem rises in the context of political decision making where epistemic discussions are misidentified as discussions about truth claims. As a result inadmissible transformations occur from plausibilities in probabilities, from dangers in risks and from illustrative data in proof. Kollek shows that different scientific paradigms in biology define different perspectives of risk assessment. She demonstrates this in the case of genetic engineering where the risk assessment model of the molecular biologist competes with the model of the ecologist. Kollek makes the point that the choice for models of risk assessment involves ethical considerations. Shrader-Frechette analyses the nature of probabilities. She uses, among others, examples from nuclear accident probabilities. She argues that the probabilistic explanation of risk assessors cannot account for reasonable fears about risks among the public. In her opinion the problem is not so much that 'objective' probabilities as given by experts are value laden, but that they are not recognized as such, and not adequately dealt with in contemporary environmental risk analysis.

The existence of a persistent dissent among scientists who adopt different perspectives and who cannot adequately address the ethical questions involved, constitutes the problem of decision making under the conditions of major scientific uncertainties. In the interchange between the value spheres, science and politics, a contradiction arises: an appeal to science seems necessary (because of the complexity of the issues), but is not possible (since there is a controversy and lack of knowledge) and what is impossible, an appeal to a source which can provide authorative data, becomes necessary.

In the entanglement of this contradiction science seems to change into its counterpart: it becomes a strategic means for political action. Interest-groups can look for scientific experts who share their political objectives. Science does not provide authorative data for consensual action and science can not adequately code the problems in terms of truth. Instead the respected value sphere causes the multiplication of dissent and conflict about important policy issues, such as the control of technological risks and environmental problems.

A way out seems the quest for new concepts of science for policy. In part II, Nowotny and Funtowicz/Ravetz discuss in their respective papers two possible concepts. Nowotny was among the first authors to describe the rise of a managerial conception of science for public policy. The clear cut boundaries between science and politics are given up, but that should not lead towards a science infiltrated by politics nor in solely science based policies. The managerial conception of science asks for a widening of the 'negotiating space' between science and politics where intermediary institutions can jump in. These new institutions, like offices for technology assessment, should allow for the legitimate definition of the political and scientific dimension of the issues. Funtowicz/Ravetz introduce the concept of post-normal science, as a problem solving strategy for cases where the decision stakes and systems uncertainties are high. In post-normal science the attention shifts from the indefinable level of uncertainties to the quality of the available information. The methodology of post-normal science will require "extended peer communities", since the quality assurance of its procedures and results involves participants outside the traditional peer communities of experts.

In part III, the use of social science in the policy making process will be explored. Pröpper gives an account of a research project investigating the exercise of power and the quality of the argumentation in evaluation research and in the utilization of this evaluation research in the policy-making process. A model procedure for discussions allows for the detection of an illegitimate exercise of power in the policy process, in the case of an appeal to data provided by the social sciences. One of his major conclusions is that the quality of the argumentation of the investigated evaluation research into the Investment Subsidies Act in the Netherlands in the period 1986-1989 influences the quality of the argumentation in the utilization of this research in the policy-making process.

In part IV the value sphere morality will be discussed, with special attention to the interchange with science and politics. In modernity scientific progress has become questionable. Science seemed independent from ethical and moral discourse. Many shared the perception that science and morality are independent. However, it always sounded paradoxical that scientific progress can be acknowledged without appealing to substantial normative presuppositions.

Kettner claims that discourse ethics, as developed by Apel and Habermas, can provide a critical yardstick for evaluating processes of consensus formation on policy issues. The mandated use of science in the policy making process is criticized for its questionable use as a substitute for a morally required consensus. He discusses some moral constraints under which the interchange of science and politics could develop. Borrillo reflects on the epidemic management of AIDS under the conditions of the rule of law in modern society. Here we see that the interchange between science and morality has become problematic. He illustrates this with the fact that doctors have become moral agents. The situation could become dramatic if we are inclined to attack the AIDS victims rather than fight the virus itself. Hohlfeld shows in his paper that the definition of disease using the biomedical model in science excludes the human ethical dimensions of disease, so that we can hardly expect that the biotechnological development in science will cover actual health needs. He makes a case for the use of concepts in science that include these human dimensions.

In part V the public and state interests in the development and control of technology are addressed. Diettrich explains the view that the ongoing conflict between industry, public authorities and other interest groups on the evaluation of risks and benefits of biotechnology can not be reduced only to a matter of missing factual scientific, legal or economic information. Analyses conducted inside and outside the European Commission services suggest that there is not that clear correlation between knowledge and attitude towards biotechnology assumed by classical public information concepts. The effect of information and knowledge depends mainly on how they are interpreted or selected by pre-existing attitudes, rather than on the factual content itself. Recent opinion surveys and other analyses launched by the Commission of the European Communities have shown that this phenomenon is strongly cultural dependent and, therefore, will vary from country to country. The question arises as to what extent scientific information can influence existing attitudes. This means, according to Diettrich that a public, industrial or R & D public information policy, aiming at a better mutual relation, has to put emphasis on both aspects: The improvement of existing attitudes, and the efforts for improving factual knowledge. Schenkelaars discusses the controversies on food biotechnology. He makes a case for a active public participation in the decision making process.

PART I

Controversies about Risks and Policy Implications

1 CONTROVERSIES AND POLITICAL DECISION MAKING

René von Schomberg
University of Twente, The Netherlands

1.1 Introduction

1.1.1 Science: a functional authority?

The relation between science and governmental policy is one of functional authority. Whenever policy makers rely on the data of scientific research they admit their belief in a scientific system which has produced that information. This belief or trust in an authority is functional since it is not founded on the veracity and blessed knowledge of individual research workers but on the competency of a scientific system (Luhmann, 1973). This trust in the transfer of truth by science is based on a functional necessity. The complexity of certain questions and problems is so enormous that we have to depend on an intermediate party for the production of information. Such information can only be utilized in a meaningful way when available in a reduced and simplified form fit for practical purposes. The scientific system can fulfil a social function since science is acknowledged as being a reliable source of supplying and production of information. The appeal to an authority depicts this function. We do appeal to an authority in those cases where we lack the ability, or do not have at our disposal the means, to verify particular statements. In such cases social actions are freed from discussions. In situations where we are forced to act without delay we can make the right decisions because of our trust in an authority. We assume the truth of the statements of an authority and we orientate our actions in accordance with the truth of such statements. Social actions are, therefore, to a large extent made possible through a trust in such an authority. The belief or trust in the assignment of truth by persons and institutions could perhaps be regarded as a condition for the development and progress of any society. With the increase in the complexity of a society the necessity also increases, in any case for particular branches of knowledge, to accept such a trust in the knowledge of other persons and institutions.

In general we may say that an appeal to the authority of science is without problems. A policy maker will not, in most cases, have doubts about the expertise of the experts in question and will, therefore, see no reason to have the available research verified. This just about always happens when the scientific information available is non controversial. If the scientific information is of a non-controversial nature, there is clarity about what should be regarded as relevant information. It is, in addition, clear that the information is potentially available. This means that the data could, more or less, be anticipated: we know what conclusions could be inferred from what sort of data. Scientific research of a non-controversial nature has a pronounced problem solving character. Since the theories and methods deployed are not in question. The problems are regarded as solvable in principle. In this type of research theories are not tested; the person responsible for the research, however, is open to test. It is the scientist who has to sort out something. The point is that he/she has to bring to light the relevant data. If he/she is successful, the mistake/fault is not supposed to lie in a wrong theory; it is the scientist who is at fault. It is the scientist who has failed in the execution of his task. The ultimate differences of opinion amongst scientists concerning the interpretation of the research data can be reconstructed in terms of a *theoretical empirical discussion*. The arguments brought forward in these types of discussions have, in principle, a consensus enforcing power. In this type of discussion we normally use explanations and predictions. In such a discussion the truth of statements are at stake. For instance, when we are discussing the predictions of a weather forecast, discussions which indeed can become controversial, the consensus enforcing power of explanations and predictions are derived from the fact that we *know* under what kind of conditions we could accept the statement 'it will probably rain tomorrow in Holland'. For example, we could accept the prediction when somebody can explain that there is a depression above the Canal heading west. The simple point I want to make about this type of discussion is that the consensus enforcing potential of predictions and explanations are derived from the possibility to explicate the truth-conditions of the controversial statements. In the case of science as a functional authority we entrust the scientific system to clarify these conditions.

The idea of science sketched above, i.e. science as a functional authority, is however not always unproblematic. In this article I will deal with one type of circumstance which is relevant in order to revise the idea of the functional authority of science. This refers to instances of scientific controversies. An analyses with the means of argumentationtheory will show that we cannot expect an agreement among opposing scientist under circumstances I am going to specify. Actually, it can be very unreasonable to demand such an agreement. I will make a proposal how to achieve a reasonable consent in the policy making process against the background of scientific controversies.

1.1.2 Scientific controversies

The functional authority of science is threatened whenever there are signs of a controversy. The belief that transfer of truth is a function of the scientific institution cannot be founded on conflicting truths of different associations of experts. Discussions in science usually unfolds around the question of how new knowledge could be obtained. In this context the available knowledge is controversial. This means that the certainty of previously accepted knowledge is questioned or regarded as limited. In such cases discussions in science could also include disputes about the methods to be employed. It is, moreover, not always clear what scientific discipline should lay a claim as to the best solution of the problem in question. Discussions about acquisition of new knowledge could be translated in terms of an *epistemic discussion*. The opposing scientist can only draw on arguments like analogies, authoritive appeals to scientific principles and counterfactual arguments which articulate uncertain and inadequate knowledge: plausibility[1]. The arguments brought forward in such discussions have no consensus enforcing potential because the experts are not able to explicate the truth conditions of statements because of a lack of knowledge. In an epistemic discussion, the controversy is not bound to the level of statement and counter-statement. Here we have to deal with a situation where a whole domain of knowledge, and even the methods of how we gain new knowledge, becomes controversial. In that case we will have to discuss the plausibility of theories and hypotheses with which the knowability of certain knowledge domains can be stated. The typical arguments in an epistemic debate do not serve the consent-achieving process but rather cope with the adequate disclosure of new domains of knowledge by constructing coherent theories, suppositions and hypothesis. In such a discussion, it is not the truth of statements which is at stake, but rather the *plausibility of knowledge-claims*[2]. Within an epistemic debate, which is about to be cleared, we can only expect a reasonable dissent but not a reasonable consent. The conflicting knowledge claims of the experts constitute epistemic uncertainty. Decisions within the field of policy making realised against the background of such discussions will, therefore, be subjected to these conditions of uncertainty. The uncertainties of epistemic discussions can lead to an inducement for a public debate. Epistemic discussions do not necessarily become public. However no one will be surprised that it is this type of debate which will have societal consequences whenever the quest for new knowledge depends on important social choices. It is more obvious, when one thinks of the increasing critical public awareness of developments within science. I will talk about a scientific controversy if epistemic discussions induce public debate.

I will start with a discussion of the structure of epistemic discussions. This I will do with the help of an example: the risks of the deliberate release of genetically engineered organisms. Since the middle of the eighties this induced both an academic and a public debate which are still going on[3] (Von Schomberg 1992b). I will not make a complete reconstruction of this debate. It is my first intention to clear the general problems of epistemic discussions. However, my presentation should clarify that all possible factual arguments can actually be covered by the concept of an epistemic discussion which I will explicate. The concept of an epistemic discussion can be utilized in different debates. This explication will be followed by some ethical questions and problems of legitimacy which arise in the context of scientific controversies. It can be shown that these problems can not be adequately dealt with in the usual policy procedures. Finally I will propose a solution which is inspired by the framework of the so called 'discourse ethics'(Habermas, 1983; Apel, 1988)

1.2 The epistemic debate about the ecological effects of the deliberate release of genetically engineered organisms

Epistemic discussions features a multidisciplinary aspect. In this case we have to deal with the competing claims of molecular biologists and ecologists. It discloses two perspectives on the new knowledge-domain of the ecological effects of the deliberate release of genetically engineered organisms.

1.2.1 Authoritive appeals to scientific principles versus analogies and counterfactual arguments

Both scientific disciplines use analogies in order to disclose the new scientific field. The different analogies lead to different assessments. Both analogies are mutually rejected on the grounds of an authoritive appeal to a scientific principle which disqualifies the experience brought in by a particular discipline. The scientific principle itself counts as undisputable within the boundaries of each discipline.

The analogy of the ecologist runs as follows:

> If one wants to judge the risks of spread of engineered organisms, one has to evaluate the chance in how far organisms can leave their predestined paths and effect the structure and function of a ecosystem. The introduction of exotic(problem) plants provides a basis for such an evaluation. In both cases naturally limiting factors for the spread and establishment are overcome by organisms.
>
> *knowledge-claim: so the experience with exotic plants provides a basis for the evaluation of the risks of spread and establishment of engineered organisms.*

The molecular biologist counters with an authoritive appeal to a scientific principle:

> One reason why critics urge caution over the release of genetically engineered plants is experience with problem plant species. By means of genetic engineering, in contrast, the organism, rather than the environment, is changed; the problems do not originate from changes in the genetic make-up of the plant but from introduction into a new environment.
>
> *knowledge claim: the experience with problem-plants is not relevant.*

The molecular biologist uses a different analogy:

> Predictions of the risk of deliberate release can be based on the experience of traditional practices in agriculture (like plant breeding). In traditional plant breeding the exact genetic changes are unknown. In the case of genetic experimentation, however, the specific modifications can be characterized. Plants have been crossed (traditional genetic engineering) by man for cen-

turies. New variants resulting from such breeding have not caused serious problems. Some crosses include those that would not occur without man's intervention. Breeders have never taken and do not now take special precautions in testing these plants in the field because they know from experience that these extensive mixings have not produced serious problems.

knowledge claim: there is no reason to expect that engineered organisms could cause greater problems than traditional techniques.

The ecologist, on his turn, can now bring in an authoritive appeal to a scientific principle:

The ecological consequences of the introduction of engineered organisms can not be predicted only with the knowledge of the genetic structure of an organism (= rejection of a scientific principle). Therefore one needs to know the biological properties of the organisms and their chance to reproduce and survive in the environment (relevant scientific principle)

knowledge claim: the knowledge of the genetic structure only can not be the basis for a successful prediction

The arguments used by the experts provide the means for an open ended discussion. The methodological analogy of the molecular biologist is encountered by an appeal to a scientific principle of the ecologist. Vice versa, the methodological analogy of the ecologist is also rejected by an appeal to a scientific principle of the molecular biologist. These arguments do not have any consensus enforcing power. Actually, one could as a non-participating scientist agree with *both* the plausibility claims of ecologist *and* the plausibility claims of the molecular biologist[4]. In the actual debate, this insight of the argumentation theorist that the plausibility claims do not *effect* each other, will not be articulated since the experts can only demonstrate their loyalty and adherence to their specific discipline.

It is not a coincidence that we have to deal with authoritive appeals and analogies in the context of epistemic discussions since they have a specific argumentative function. In the use of analogies one makes a case for methodology. The molecular biologists say that the risk issue should be studied in terms of the genetic characteristics of an organism. The ecologist maintain that one should study the biological properties of a organism. The different analogies refer to the different methodologies of scientific disciplines. These methodologies are indisputable within the boundaries of the disciplines, but are not self-evident for disciplines which claim knowledge of the same issue. On the contrary, in a large number of practical debates one rejects the opponent's view as 'unscientific'. In a dispute about methodology, there is no discussion about the truth of statements; there are, however, different claims as to how new knowledge should be acquired. The different disciplines develop their own perspective on this problem. The use of an appeal to a scientific principle is a comparable case. The 'scientific principles' are, again within the boundaries of a discipline, acknowledged as reliable sources, but as one transgresses the field they can become controversial.

In the use of analogies we mobilize knowledge from well-known areas of research. In our case the ecologists do not have the knowledge in order to predict the ecological consequences of the introduction of engineered organisms. Therefore they try to mobilize knowledge (using an analogy) from the field of the introduction of exotic plants. Such an analogy enables ecologists to constitute a domain of possible relevant facts. In this way analogies have the function of the mobilization of knowledge. In the use of an appeal to a scientific principle we are also confronted with the possibility to claim knowledge. In this case an appeal to a principle enables us to get access to a certain problem. In complex scientific issues we are confronted with a whole domain of inconsistent and incoherent data which cannot all be assessed. An appeal to a scientific principle can reduce the complexity of the issue. In this way we can make an issue suitable for research. So analogies and authoritive appeals state the possibility to do research in new domains and to anticipate the relevance of data which are still to be gained.

We have seen that the arguments put forward do not actually effect each other but only articulate the epistemic uncertainty of the new field. A counterfactual argument is the only means at our disposal, to doubt the plausibility of a claim. This argumention form can force us, sometimes in the form of a reductio ad absurdum, to reconsider the premises of our arguments. In an epistemic discussion the molecular biologists make, for example, instance an appeal to the principle of adaptation in evolution-theory: "Pre-existing organisms compete successfully with genetic engineered organisms in the environment because the former are better adapted to the environment". The ecologist can challenge the plausibility of the claim of the molecular biologist with a counterfactual argument:

Suppose: it is correct that pre-existing organisms compete successfully
it follows: if pre-existing would compete successfully then it would be impossible for engineered organisms to persist
but: genetic engineered organisms are designed to have a function in the field and therefore they obviously can not be out competed too soon

so: it is plausible that for genetic engineered organisms to be of any use, they must at least persist to some extent.

It may, in the case of counterfactual argument, be easier to demonstrate that we have to deal with plausibility claims. It does not make sense to judge the conclusions of a counterfactual argument in terms of truth conditions. In a counterfactual we do not start with a *proposition* but with a *presupposition* which is announced by: 'let us suppose that' or: 'suppose for the sake of the argument that.' In a counterfactual argument we could even start with a premise which involves the knowledge-claim of a whole theory: let us presume this theory is true. In the case of epistemic uncertainty, we are most likely to encounter this kind of argument. The imagination of experts, oriented to (subjunctive) thinking of what might happen, is challenged by the unknowns of the issue. Plausible arguments, which are neither inductive nor deductive, do not feature qualifiers[5]. Qualifiers normally, depict the conditional character of truth claims. Yet, in the context of plausible reasoning they do not make sense. We cannot say that something is *presumably* plausible. We can neither say that something is *generally* nor that something is *obviously* plausible[6]. These remarks reinforce our intuition that 'plausible' unlike true is not a special predicate of statements but rather of knowledge-claims.

In epistemic discussions we have to make use of the weak arguments mentioned above. They are more or less indicators for a fundamental lack of knowledge. They all have a function in the acquisition of knowledge, especially in the context of controversial knowledge-claims. It is important to notice that these arguments can not establish conclusions. They are brought forward to make plausible and promising proposals, still in need of further investigation.

1.2.2 Prospective plausibility claims

On the basis of analogies and authorative appeals to scientific principles expectations having the appearance of predictions can be made. Molecular biologists; e.g., appealing to the principle of adaptation in the theory of evolution claim, for instance, prospective plausibility for the assertion that genetically engineered organisms when released in the environment cannot overcome the pressure of selection of already existing organisms. Ecologists, however, maintain that on the basis of their experience with the introduction of new species of plants we have to consider similar unforeseeable catastrophes as had occurred with the introduction of exotic plants. The claim to predictions (in the theoretical scientific sense of the word) can only be made under conditions of presumed valid knowledge. This, however, is not the case since only plausible principles are available which still have to be found acceptable in this new field. This important epistemelogical difference has real pragmatic consequences. The corroboration of predictions can be seen as a test of the law-like statements involved, and, would then count as an element of a discourse about truth. On the contrary incomparable prospective plausibility claims can in the long run only be qualified as pragmatic selfconfirmations of their own special paradigms. The competing plausibility claims of course refers to the same subject matter but cannot be related directly to each other, and, what is more, the non- confirmation of a prospective plausibility claim can have no falsificationary value[7]. According to my opinion the examples as well as facts incidentally introduced in arguments and having an illustrative function only, should be seen in a similar way. They have no falsificationary power they rather justify the attempt to investigate the same subject matter from different authoritative points of view of the different scientific disciplines in question. Figure 1 represents a view of the structure of an epistemic discourse.

1.2.3 The definition of problems by the experts and the inconsistent recommendations to the public and the political arena

Ecologists and biotechnologists put forward different recommendations to politics in accordance with their particular scientific argumentation. Biotechnologists advise that a no particular regulation of the new technique is necessary: the existing control of the traditional methods for plantbreeding guarantees sufficient certainty given the analogy with natural and traditional processes for breeding. The advice of ecologists is found by their argument on the data that a prediction as regards the ecological effects of a release into the environment is impossible in principle: new tests should be developed to investigate the products of genetic technology with respect to the case in question. Both scientific groups demand a 'science regulated policy' but put forward unavoidable incompatible assessments of the problem at hand[8]. Ecologists define the problem as being a new problem while biotechnologists maintain their own principle everything always remains as it has been.

Figure 1 The structure of epistemic discussions

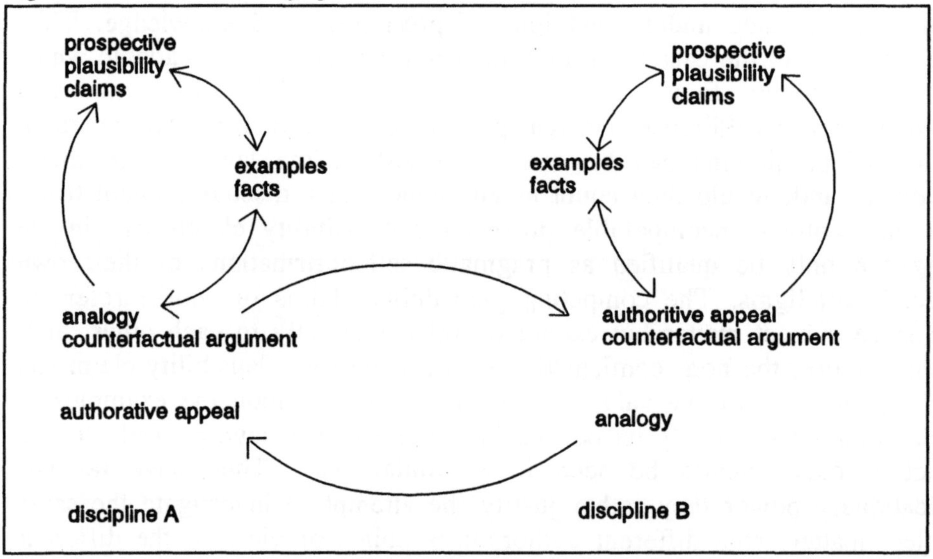

Moreover the biotechnologists propose that decisions be made in accordance with available information whereas ecologists believe that a new definite decision can only be made after new knowledge has been acquired.

1.3 Epistemic discussions within the context of contemporary political decision making

In justifications of decisions in policymaking we without any doubt mostly employ a form of argumentation which could be reconstructed in terms of the well-known model of rational decision making. The idea of scientific agreement as the foundation of a rational policy could be found in the model of rational decision making which e.g. forms the basis of cost-benefit analyses. In this model we start with given problems where the policy maker has at his disposal clearly defined aims of policy. Besides this, he also has available a list of alternative means whereby each of these aims could be realised. His choice of means is based on specified rules of preference and selection. Scientific information occupies an important position in this model since it determines the possibility of realizing the aims of a policy and the effectiveness of the means. The quality of a policy is expressed in terms of effectivity and efficiency. The ideal conception, forming the basis of that rational model of decision, rests on the assumption that an agreement (consensus) amongst scientific experts guarantees the quality of the policy.

These important assumptions cannot be made when we would have to make an appeal to information borrowed from an epistemic discussion in science. Against the background of epistemic discussions there appear the uncertainty of our present knowledge: the principal incompleteness of the information and the inconsistency of the data. The scientific information can now constitute a strategic source which, dependent on the different political options, could be interpreted differently. Within this context of the contemporary political decisionmaking and the epistemic discussions we can find the following phenomena whereby the inadequate rationality of societal actions could be catalogued.

1. An inadmissible translation of the data from an epistemic discussion to expressions having truth functional or probability characteristics (or the translation of illustrative data in proof, or the transformation of dangers in risks)[9]. This can be seen, e.g., in the case of the translation of plausible knowledge claims to predictions with probability characteristics. Such 'probabilities' must then, again and again, be adjusted to suit new events, states of affairs, and catastrophes. This insight is generally acknowledged

after the Chernobyl event. In our example of genetic engineering the data of field experiments have to prove the safety of genetic engineering.

2. In a particular policy preference is given to a discipline that can provide facts (data). The hypothetical dangers formulated by opposing scientific disciplines do not sufficiently flow into the process of decision making[10]. The emphasis on the inadequacy of our knowledge constitutes a loss of scientific authority for that particular discipline rather than a problem for political decision making. The discipline which accords with the subjective preference of the policymaker is the winner (decisionism). In Europe we come across the remarkable situation that what is dangerous in Denmark (a stop (moratorium) on the deliberate intentional freeing of genetically manipulation organisms) is regarded as inquestionable in Italy (the absence of any control; at least depending on an eventual EC control).

3. When an epistemic discussion attracts the attention of the general public the result is a quite often an unproper politicising effect on scientific debate. This can be shown on the one hand in the irrational struggle concerning the data: interest groups look for support from experts who share their political objectives. On the other hand one reacts affirmatively on the presuppositions of the model of rational decison making[11]: as long as the controversy continues, there will be no decision. Depending on the political preferences one feeds a controversy with explosive new data or one tries to escape scientific dissent. It can also be the case that interestgroups call for political decisions, whereas in the policy making process one refers to a controversy in order to delay decisions. I do not want to reinforce the wide accepted notion that in the case of controversies, interests come into play. I rather want to stress a point what should be important for a theory of society, the fact that science has become a resource for strategic action. It looses its functional authority. This means that science looses its function of unburdening societal actions from (unnecessary) discussions. In the policy making process a contradiction arises: an appeal to science seems necessary (because of the complexity of the issue), but is not possible (since there is a controversy) and what is impossible (an appeal to a source which can provide authoritive data) becomes necessary.

4. The sequence of the actions in the policy: collecting the facts before making the decisions leads to a situation where scientific unsolved or

unsolvable problems do not appear on the agenda. Science is put under pressure to produce hard facts and, that, in those fields where we cannot expect them to do that. What is wanted are 'one handed scientists'(Rip,Groenewegen 1988, p 149-172)

5. The scientific debate amongst disciplines for claiming acknowledgement and authority as regards a new field of research is threatened to deteriorate into public campaigns for recruiting declarations of sympathy. Biotechnologists promulgates the promises and blessings of the new technology while ecologists do the same for a threatened environment.

6. Even the legal system seems to be unable to cope with the problems. This is apparent from the following:

a. The principle of the causal agent (blame) cannot be applied. Eventual "actors" nor "victims" can be identified when using the new technology. It is, e.g., impossible for a victim of the Chernobyl disaster in Europe to go to a court of law and claim that the disease from which he is suffering has been caused by nuclear radiation.

b. Legal norms can no longer be controlled in practice. (The observance of standards is often the result of informal agreements (and negotiations) between public authorities and individuals. In some cases legal norms cannot even be defined. In most of the western countries the admissibility of the maximum amount of radioactive radiation in the vicinity of nuclear reactors is determined by what laws concerned with atomic affairs refer to in terms of 'the most recent state of affairs in science and technology'.

c. The legal system can no longer fulfil a normative role as regards the admissibility of technological actions. this is very well illustrated in a judgement of the "Bundesverfassungsgericht"(supreme court) in Germany concerning the controversy about the Kalkar plant: "It is not the duty of the lawgiver to determine the possible kinds of risks, factors of risk, the methods to determine such, or, fix the limits of toleration". It is obvious that the judge has shifted this problem to the realm of politics(Wolf, 1991).

d. The conflicts in society cannot be settled under the conditions of the equality of power of a judicial judgement; it is left to a social unequal power struggle where human beings depend on the responsibility of individual citizens. This is a difficult and in the future undoubtly unrealizable task due to the fact that the risks of the new technologies can no longer be observed by the individual citizen.

Against the background of these phenomena there arise the problems of legitimacy confronting the planning state who, on the one hand, can no longer agree with the definitions of problems of interestgroups, and, on the other hand, does not know how to cope with the disapproval of the overpowering processes of innovation about which the citizens cannot make decisions. The problems of legitimacy is partly compensated by the tendency of public management to negotiate with different groups in society. Examples are to found in the representation of such groups in the councils of health and environment. I do not expect spectacular results from mutual concessions reached during the negotiations since such concessions normally arise wtihin the framework of strategic actions and unequal conditions of power. It is far more important to note that on this road the question concerning the way to solve by means of a justifiable procedure the historical new problem of making decisions under conditions of scientific uncertainty is abandoned. The fact that we do need a procedural solution is apparent since there does not yet exist an accepted institution in society that could determine which actors in what way would participate in the decisionmaking within the context of scientific controversies. In the last paragraph I will try to show that the use of a procedural process applied to such problems would generate the general framework for a justifiable solution. In this way the problematic phenomena (mentioned under 1 and 6) could be eliminated.

1.4 A discursive procedure under the conditions of administrative law

The socalled "discourse ethics" could provide an understanding of the way to answer questions about just procedural solutions without getting stuck in ethical partial criticism of technology (from abortion to nuclear energy) or the dogmatic unwillingness to make certain values subject of an argumentative test. Within discourse ethics one argues that a material ethics maintaining that one could, for once and always, prescribe norms cannot be founded on arguments. History has taught that our ethical and scientific insights have, time and again, been shown to be fallible, and, that those insights had to be revised in the light of new situations and problems. The validity of norms can no longer be derived from sources that have been regarded as infallible. The validity of norms should rather be sought for in the free mutual acknowledgement of these norms by (potential) discussion partners as this could only be found in discussions. The conditions for the acknowledgement of norms is, therefore, of the utmost importance (as a matter of fact this is true for all procedural theories of justice and democracy). In one particular sense we could put the case that the conditions for a discussion are simultaneously the conditions for a rational agreement (about norms). In the light of our question about the way to reach an agreement about policy, given the background of epistemic discussions, we can make a list of some conditions which have to be fulfilled in discursive procedures.

1. In the analysis of the structure of epistemic discussions we have to establish the idea that there should be an acknowledgement of scientific disagreement. It is not reasonable to expect that scientific experts on the basis of scientific insights will be in agreement in the near future. This means that in procedures concerned with policy one sided participation of scientific experts cannot be justified.

2. The problem of legitimacy cannot be solved if some of the possible parties concerned with the politic process are excluded. Another condition is that all parties should have equivalent roles[12]. Experts should therefore only be allowed to supply information for the discussion and should not have an advisory or determinative function as regards policy.

3. Decisions having irreversible consequences for non-identifiable groups or future generations who are not able to participate in the discussions should carry the burden of a heavy legitimacy and should, if possible, be evaded.

Discursive procedures implies that those norms will be rejected which seems to exclude the universalisable interests in anticipation. In discursive procedures the norms are not grounded, but selected negatively.

4. All relevant aspects of the problem should be dealt with within the framework of a discursive procedure. Next to the eventual scientific problems field the following are also relevant:
ethical questions such as: what options are desirable? (in general: how do we want to live?)
moral justification such as: what norms should in the interest of all be included (e.g. questions about the division of risks in the society)
questions about justice: What (social and technological) aims should be promoted or limited within the framework of the rules of law. (Example: should biotechnological research be controlled by legal means?)

During the short history of technological policy in the Western countries we find a few proposals to develop procedures of this problemfield. During the seventies there were heated discussions about a Science Court (especially in the United States). The idea being an occasional body where scientists from different fields could act as judges in order to reach agreement, however, with the promise that normative aspects should be kept out of the discussions. This idea however was never realised in an institution.

In the light of our foregoing analysis such a body would not be able to contribute towards the solution of the controversies since the procedure in question is based on the assumption of an agreement amongst scientists if the normative aspects are set aside. That this was an illusion became evident very soon. Another form of developing procedures, however, was institutionalized. This took the form of the so called Technology Assessment. Initially this implied the establishment of an instrument of planning where the expected effects and side-effects of technology could be mapped and used as input in the process of making a decision. In this case the possibility of a rational consideration of the pro's and con's of technology was the guiding light. An Office of Technology Assessment was set up as early as 1973 in the United States. This office is an advisory board for the American Congress. In some European countries there exist similar offices; in the Netherlands, e.g., since the middle of the eighties. That office can give advice to the Minister either voluntary or by request. This development of a procedure could be seen as the first attempt at institutionalizing since it aimed at an actual and democratic guidance and control of technology. This essential element, however, hardly manifests itself in the real functioning of these offices. This is not possible due to the fact that an 'evaluation' of technology always comes to late. The public information on new technologies only starts moving when the products of the new technologies have been realized. Moreover, the information on the new products are, quite often, too limited by patent laws. The six phenomena mentioned under 1.3 do not seem to be eliminated by the presence of Asessmentoffices (certainly not in their present form). What is more, there is no acknowledgement (at least not of the nature of a procedural acknowledgement) of any scientific uncertainty or scientific dissent. Assessmentoffices seem, as far as a commissions for interdisciplinary research are concerned, to anticipate agreement amongst scientists. Epistemic discussions are, ultimately, analyzed in terms of conflicting interests. This leads to the loss of the possiblilty to select on the basis of universalisable interests. Not with standing these remarks, the Dutch office, for example, has contributed towards a social learning process which could develop in the direction of a discursive procedure. From this point of view the office has a social function. Social groups could approach the office to make known their desires and need of information. It would, however, be an essential improvement if all the discussion parties involved could be granted equal power in a discursive procedure. In order to achieve this, however, the necessary change as regards administrative law has to be introduced. The conditions for a discursive procedure should be legally settled and, above all, the rights and duties as regards the distribution of information should be installed. The policy process would then no longer be of a evaluative nature; it will become constructive in the form of a continuous interaction between, the providing of information, exchange of information and determination of policy. In such a policy process real democratically controlled learning processes with technology could be implemented, - and last but not least - mistakes could be restored. In any case it is plausible that the phenomena mentioned under 1.3 could be eliminated.

Notes

1. Epistemic plausibility is, on the one hand, related to plausibility on plausible arguments which are neither deductive nor inductive. They do not answer to the traditional formal-logical claims. On the other hand epistemic plausibility is related to the plausibility of assumptions and premises having a non-propositional structure(see also Rescher 1976). The plausibility of conclusions does not rest on the presupposed truth of the premises, but acquires its authority from the reliability of the sources of knowledge to which can be appealed. The plausible arguments referred to, by me, are up to now as far as I know the only explitication of Rescher's insight.

2. Many problems in the traditional philosophical theories about truth are, to my mind, founded on a confusion between reference to truth and reference to experience; or, to put it in modern terms to the confusion of discourse- theoretical truth and epistemic plausibility. Peirce who explicated the concept of plausibility , offered a starting point for a solution. Those authors who work in the tradition of Peirce, however, make either an absolute claim of epistemic plausibility (Rescher) or an absolute claim of discourse theoretical truth (Habermas,1973).

3. See especially: Brill, W.J., *Science*, vol 227, 1985, 381-384; Colwell, R.K., et al., *Science*, vol 229, 1985, page 111 and Davis, B., *Science* vol 235, 1987, page 1329.

4. This would, of course, not be the case where it concerns conflicting claims on truth. Accepting a truth-claim does require the refusal of a conflicting claim. The plausibility of knowledge-claims, however, is only touched by the paradigmatic internal coherence of particular assumptions and statements. There is a class of theoretical linguistic differences which cannot be explicated here. Intuitively, it should be clear by now that plausibility and truth do not have to converge.

5. This includes the abductive conclusion, not mentioned here.

6. This idea is missing by Toulmin (1958,1984)

7. The claim that the structure of explanations and predictions are identical has been dropped after a debate which lasted for two decennia (see H. Lenk (1986). Stegmüller (1969) already differentiated a list of 30 different types of predictions. The structure of prospective plausibility claims has not been revealed up to now.

8. Occasional attempts have been made at a solution in the form of interdisciplinary research. The experts, then, argue unendlessly about which disciplines should participate in the research.

9. For the transformation of dangers in risks see Evers, Nowotny (1987).

10. In the policy making process one cannot deal with the concept of hypothetical risk either. See Kollek (1988, p 34).

11. I do not mean that one, in the political realm, actually explicitly turns to the normative model of rational decision making. I only assert that the empirically founded arguments in the policy making process could be optimally represented in this way.

12. (Kettner has explicated a number of other conditions, see chapter 7 this volume).

References

Apel, Karl- Otto (1988)
Diskurs und Verantwortung, Frankfurt am Main: Suhrkamp

Evers, Adelbert/Helga Nowotny (1987)
Über den Umgang mit Unsicherheit. Die Entdeckung der Gestaltbarkeit von Gesellschaft, Frankfurt am Main: Suhrkamp

Habermas, Jürgen (1973)
"Wahrheitstheorien", in: H. Fahrenbach (hg.), *Wirklichkeit und Reflexion*, Pfullingen, p. 211.

Habermas, Jurgen (1983)
Moralbewusstsein und kommunikatives Handeln, Frankfurt am Main: Suhrkamp

Kollek, Regine (1988)
"Verrückte Gene. Die inhärenten Risiken der Gentechnologie und der Difizite der Risikodebatte". in: *Ästhetik und Kommunikation* nr. 69. p. 29-39

Lenk, Hans (1986)
Zwischen Wissenschaftstheorie und Sozialwissenschaft, Frankfurt am Main: Suhrkamp

Luhmann, Nicolas (1973)
Vertrauen, Stuttgart.

Pimental, Donald. (1987)
"Down on the Farm: Genetic Egineering meets Ecology". in : *Technology Review*, nr. 1, p. 24-30

Rescher, Nicolas (1976)
Plausible Reasoning. An Introduction to the Theory and Practice of Plausibilistic Inference, Assen: Van Gorcum

Rip, Arie/Peter Groenewegen (1988)
"Les faits scientifique a l'epreuve de la publique", in: Michel Callon (Hg.), *La Science et ses réseaux. Genèse et circulation des faits scientifique*, Paris/Strasburg, p. 149-172

Schomberg, Rene von (1992a)
Argumentatie in de kontext van wetenschappelijke controversen, Enschede: Fakultaire reeks van de Fakulteit voor Wijsbegeerte en Maatschappijwetenschappen Universiteit Twente.

Schomberg, Rene von (1992b)
Die Darstellung des Verhältnisses von wissenschaftlichen, wissenschaftspolitischen und politischen Argumenten in der Debatte über die Freisetzung genetisch manipulierter Organismen. *Paper des Wissenschaftszentrums Berlin.*

Stegmüller, Wolfgang (1969)
Wissenschafliche Erklärung und Begründung. Bd.1 der *Probleme und Resultate der Wissenschaftstheorie und analytischen Philosophie*, Berlin-Heidelberg-New York: Springer

Toulmin, Stephen (1958)
The Uses of Argument, Cambridge: Unversity Press-/Richard Rieke/Allan Janik (1979), An *Introduction to Reasoning*, New York: McMillan

Wolf, Rainer (1991)
Zur Antiqiertheits des Rechts in der Risikogesellschaft, in: Beck (hrsg), *Politik in der Risikogesellschaft*, Frankfurt am Main: Suhrkamp

2 CONTROVERSIES ABOUT RISKS AND THEIR RELATION TO DIFFERENT PARADIGMS IN BIOLOGICAL RESEARCH

Regine Kollek
Institute for Social Research, Hamburg

2.1 Introduction

Science-based complex technical processes and/or their products like radioactive isotopes, halogenated carbohydrates or genetically engineered organisms are often characterized by the fact that their specific risks may involve potentially disastrous and/or irreversible consequences. Another feature of such technological systems is that there are no clear cut boundaries between research and technological development, since development becomes a scientific strategy, which extends science beyond the laboratory. The applications of such technologies or their products therefore have an experimental character, which makes it on the one side impossible to accurately estimate the probability of failure of such systems and their safeguards on theoretical grounds. On the other side, due to the presumed low probability of failure it could be impossible to empirically confirm preliminary estimates in practice. (Weinberg 1972) Moreover, because of the high consequences of failure, testing by trial and error of such technologies poses an inacceptable high risk to the human population and to the environment. That is why this type of risks was called "hypothetical" (Häfele 1974), in contrast to the "empirical" risks of more traditional technologies, which can be verified without inacceptable consequences.

To date, the discussion about hypothetical risks has focused on enhancing the calculation of probabilities, on questions of containment and on the acceptability of long-term impacts or so-called "residual risks" (Restrisiken). However, this approach does not take into account that although probability assessments are essential to making engineering decisions about competing designs or alternative materials, they are not the only factor which influence societal technology choices. In fact, they may even be largely irrelevant. (Rayner, Cantor 1987) Furthermore, emphasising probabilities may obstruct the very specific characteristics of hypothetical risks. An understanding of these characteristics is needed however, in order to make rational decisions about the acceptability of technologies which carry this type of risks.

In this paper I will try to elaborate some of the characteristics of hypothetical risks and their intimate relationship to science based technologies. This complex relationship is approached on a very fundamental level. Starting with one of the most intensely debated fields of modern science, with molecular biology and its application in genetic engineering, the hypothesis is put forward that new uncertainties and risks are generated not only within the context of application of scientific knowledge and technological development, but rather have their roots in an earlier phase, in the "context of discovery" and the theoretical-experimental basis of such high technologies itself. Thus, my colleagues W. Bonß, R. Hohlfeld and I have proposed a junctim between the "experimental philosophy" of science and the "philosophy of risk assessment" of science-based technologies. (Bonß et al. 1990, 1992) In support of the notion that new uncertainties and risks are generated by science itself, I shall specify some aspects of genetic engineering and its epistemic background, in order to demonstrate the relationship between scientific methodology and the generation of hypothetical risks. Finally, some conclusions are drawn from this analysis for scientific an public discourse on risk perception and acceptance, and some suggestions are made, how this could be transformed into research and policy strategies, which may create new or alternative options for science and society.

2.2 Perspectives in risk assessment

By the help of molecular biology and gene technology, organisms are being developed with genetic information and phenotypic properties which they previously did not possess. It is this novelty, which is the basis of the scientific and industrial usefulness of transgenic organisms, and which makes genetic engineering and its products so interesting for a whole range of applications. And as it will be shown, it is precisely this novelty which contains the risk.

In microbiology and biotechnology, the classification of ordinary microorganisms is conducted on the basis of known levels of safety and danger for humans as well as for laboratory and agricultural organisms, reflecting longstanding empirical proof, and is not, therefore, based on theoretical considerations. This concept has also been applied to work with organisms which have been altered by means of genetic engineering. For example, the european and german regulations for the contained use of recombinant organisms and the deliberate release of such organisms require an assessment of the recombinant phenotype. But since there are only limited means available to fully assess the phenotype of such an organism without releasing it into the environment, in the actual practice of risk assessment it is often assumed that a recombinant organism does not pose a higher risk compared to the original host organism, plus the specific risk potential of the foreign gene which has been introduced. Such a classification is thus based on the addition of the characteristics of the host organism and those of the transferred gene, the so-called "additive model". (Kollek 1988a, 1988b) According to this view, the phenotypical characteristics of an organism are seen as the result of the sum of its genes. The addition of a specific gene causes, at most, the addition of the traits coded for by the transferred gene.

According to this understanding, the gene is a discrete unit and a carrier of information which is independent of the organism or the specific genetic background, that is of context-independent information. Seen from this perspective, one would not expect organisms to develop surprising or unknown traits by the transfer of genes with known nucleic acid sequences. Although these basic principles of risk assessment are accepted widely by legislative and executive bodies, for many biologists and environmentalists it is highly questionable, whether this model, based on a combination of empirical evidence and specific theoretical premises about the nature of genes, can be applied to genetically engineered organisms. To their opinion, such a model can at best, be used as a base for risk assessment in cases in which complexity is low and the interconnections of different systems and levels of complexity are limited. This is, for example, the case when manipulated organisms are used in controllable surroundings with a low level of variation, e.g. within the physical containment of a biotechnological production unit. Such a perspective becomes problematic however when complexity and interactions are high. This is the case when transgenic cells or organisms are deliberately or accidentally released into the environment, and when the biological effects of the transferred nucleic acids unfold in epigenetic and ecological contexts.

This controversy about the premises of risk assessment and their practical consequences have been one of the main issues in the public debate on genetic engineering during the last years. In order to elucidate the problems related to different perspectives and models, I want to discuss some of the terms which are of critical importance for the definition of biological risks.

2.3 Controversial concepts of the relationship between genotype and phenotype

One of the most challenging task in molecular biology is to understand the relationship between changes in the genetic material and the phenotype of the recombinant organism. Furthermore, the relationship between phenotypic properties on the one side and the occupation of a specific ecological niche on the other side needs to be known, in order to make as reliable predictions as possible about the interactions of the transgenic organism with the environment or its components. They are therefore most relevant for the description of the characteristics exhibited by transgenic organisms, and the risks associated with genetic manipulations.

An indispensable prerequisite for approaching these complex relationships successfully is the elucidation of the gene concept. Today in molecular genetics, the term "gene" refers to a piece of DNA, which can have different functions and different structures. But although such a "realistic" concept, which refers to the gene as a materially existing entity, is the most prominent in molecular biology discourse, it is not the only and probably not even the most convincing concept. It competes with an "idealistic" or "theoretical" concept, which regards the gene as an instrumental unit or an intellectual device to organize data. (Falk 1984) But even if one considers only the "realistic" concept, things are not easy at all. A gene for example can be a regulator for gene expression or be translated into a protein. Genes also can come in pieces and jump around in the genome. Although many characteristics of an organism are explained as being a result of a definable relationship between a gene or certain genes and phenotypic traits - let it be a relationship of one to one, or one to many, or many to one - many others do not follow these simple rules. For example, the expression of a gene and its effect on cell physiology may depend on their chromosomal location and their neighbours along the chromosomes. This phenomenon is called "position effect" (Sturtevant 1925), and it poses a challenge to the "discrete gene concept", which has not been solved satisfactorily to date. (Falk 1984) In spite of the elucidation of the molecular structure of many chromosome segments, knowledge of the sequence and the biochemical make-up of DNA or of a specific protein therefore does not - or at least not in all cases - allow to infer which function a particular gene or protein will have in the cell, or how the activity of that protein will affect the physiology of the respective organism, or its interaction with other organisms or environments. (Kollek 1988a, Nagl 1992) These findings support the notion that an "additive" approach is neither sufficient to predict the functions of a specific DNA segment in a cell or an organism, nor is it capable of describing fully the phenotype of a cell or an organism, or its interactions with the environment.

Without going into more details, one can say that today, there are divergent views of how the relationship between genotype, phenotype and ecological niche should be conceptualized. Risk assessment therefore is confronted with a dilemma, which can be described as follows: on the one side, the "additive model" makes it relatively easy, to categorize a certain experiment or transgenic organism into a specific risk group, by summing up the properties of the parental organism and the transferred gene. On the other side, it carries the disadvantage of not grasping specifically the phenotype of the *transgenic* organism. A "contextualistic" or ecological model however, which takes into consideration the contextual relationships between genes, and genes and environment, would require extensive research to empirically evaluate the actual phenotypical properties of the altered organism.

Whereas the second approach complicates and lengthens the risk assessment procedures in a way that is said to be unacceptable for industry, the first model leaves the interactions of the scientific object with open systems out of scope. These divergent positions reflect on the one side different policy strategies in evaluating the risks of transgenic organisms. But on the other side, they also reflect epistemic controversies. Although the problem of how to conceptualize the relationship between phenotype and genotype was reformulated in the context of genetic engineering, it is not new: it exists in the scientific discourse since the beginning of modern genetics. (Falk 1984; Levins, Lewontin 1985; Sattler 1986)

2.4 The epistemic background of experimental science and the limits of experimental knowledge

To understand the practical relevance of such epistemic controversies, it is necessary to recall the rules and procedures of experimental science, which were designed in order to optimize the process of scientific inquiry. Since many natural phenomena and events are of tremendous complexity, their underlying principles cannot be discovered by observation alone. The introduction of systematic experimentation as a methodological principle into biological research helped to overcome this problem. In the course of an experiment, objects are withdrawn from the real world complexity and examined under controlled conditions, where they can also be exposed to specific factors and influences. This "methodological reductionism", which is a basic requirement for scientific understanding, helps on the one side to describe isolated phenomena and their properties more precisely. But on the other side, it also excludes complexity and coherences and can therefore be described as a process, which enhances and restricts perception at the same time. (Bonß et al. 1992)

Reduction of complexity for the sake of a more precise recognition of isolated phenomena is not *per se* problematic, at least not as long as scientists are aware of the fact that this kind of reductionism is a prerequisite of scientific perception and therefore an implicit part of the object itself. But when experimentally altered objects (like genetically engineered organisms) are released into the environment, it has to be considered that data obtained under restrictive conditions show only those aspects, which were not excluded by the practical or theoretical presumptions of the experiment. Reductionism therefore becomes problematic, when natural phenomena are reduced to theoretical models, which select some aspects and regard them as more important than others. This kind of "theoretical reductionism" reduces the phenomena to that what is seen under the theoretical and practical premises, e.g. in the context of laboratory science, which systematically abstracts from contexts beyond the laboratory walls. During this process of "de-contextualisation", phenomena are looked at under specific "boundary conditions", which prescribe, what can be seen and how it can be seen. (Polanyi 1969)

Today these principles of experimentation are also applied to cellular and molecular biology. Following the paradigm of theoretical and methodological reduction of complex phenomena to ever more simple elements, experimental approaches in molecular biology concentrate on the elucidation of molecular mechanisms within cells and the genetic base for these mechanisms. Through the exclusion of preceding contextual relationships, objects and phenomena are stripped of seemingly superfluous, unnecessary or undesired complexity, which hinder the identification of the "real nature" of the object or process in question and the rule, by which it is governed. Thus contextual relationships themselves are not in the scope of scientific inquiry. Due to this selectivity, the rules of scientific reasoning and experimentation thus prescribe, what can be perceived by scientific means, and what has to be excluded from the experiment. Therefore, they implicitly define, what is relevant to science and what is not.

This canon of methodological rules was defined by the philosophers of the enlightment and their followers in order to rationalize scientific discourse and to confront what was perceived as "wild speculations" about natural phenomena with a systematic search for truth. Conclusions drawn by means of scientific logic and methodology were claimed to be true and objectively given. But today, we have to put this contention more precisely: by scientific methodology, phenomena are isolated from their context, and coherences are excluded. What we learn by laboratory experiments therefore does not represent knowledge about "real" nature, but rather knowledge about experimentally manipulated objects. The part of nature studied this way, has actually been re-created in an artificial world, which is structured by man-made rules. Scientific statements are thus relevant at first for that what can be grasped through scientific methodology and the technical instruments which have been employed. They do not directly apply to the behaviour of the object of study in the world outside the laboratory. Furthermore, different methods describe the object of study from different perspectives and thus produce different images of reality.

This does not mean that we can not achieve a close approximation of an understanding of reality through systematically searching and asking questions, complemented by historical and practical experience, so that we are able of building instruments and production units which function. The interpretation of science - as it is formulated here - as a strategy to acquire instrumental knowledge does not question its powerfulness, its precision or its successes with regard to the construction of new effects and products. However, it shows that hypotheses used as a starting point for the formulation of questions about nature and rules and methods which are intended to help in answering these questions, do not reflect nature as such, but emerge in specific historical, institutional and social contexts and are subject to social change. These rules structure the framework of experience and action by scientists. The structuring is effective with the help of different mechanisms, among them for example specific lines of questioning, special technical instruments, patterns of action and of perception, as well as a specific language and a body of knowledge which accumulates in the course of time. These mechanisms stabilize what Thomas Kuhn has called a paradigm, which can be defined as a system of laws which determines what kind of questions are acceptable, which strategies of answering these questions are considered scientifically sound and which are not. It represents a semantic context, which allows the interpretation of empirical phenomena, and defines a framework within which normal science takes place at a certain time. (Kuhn 1967) Theories thus are not neutral in their relationship to natural phenomena, and what we observe experimentally is influenced by the questions asked and the instruments used.

The process of de-contextualization represents a fundamental scientific principle. Defining contexts of not being relevant and using restrictive conditions is on the one hand the prerequisite for the success and efficiency of the control and manipulation of scientific objects. But on the other hand, it is also tightly linked to a loss of predictability of its behaviour in open systems. De-contextualisation therefore is intimately connected with the generation of uncertainty and risk, once the altered objects leave stringently controlled environments. In establishing this junctim between the "experimental philosophy" of science and the "philosophy of risk assessment" it is therefore argued that preexisting environmental and semantic - that means conceptual - contexts must be taken into account as a necessary, but not sufficient prerequisite for risk assessment of genetically engineered organisms and products.

On the grounds of this specific description of the very fundamental characteristics and limits of experimental knowledge, it is now possible to come to a more precise understanding of its consequences, once it is applied beyond the rationale of restricted systems. Since in the process of experimental and theoretical abstraction natural objects are stripped of their environmental and semantic contexts, concepts formulated on the basis of laboratory experiments are not sufficient to predict the complex interrelationships of these objects when they enter the world outside the laboratory and are confronted again with complexity and contingency.

Today, organisms which are genetically modified on the basis of experimental knowledge are being released into the environment. At present, their number is still relatively small, but the problem of predictability becomes more significant when large scale applications of such products takes place in the future. Released organisms may not behave as assumed on the basis of laboratory or small scale studies. In some cases, they may not be able to establish themselves in the environment and die off; in others they may cause small or large scale damage. It is questionable, whether such applications or involuntary releases for example from biotechnological production plants will be reversible in every case. Since they are done on the basis of restricted knowledge, they are in fact experiments in the environment and with the environment, as Krohn and Weyer (1989) and other authors have pointed out.

2.5 Conclusions and consequences for scientific and political discourse

The following consequences can be drawn from this analysis for scientific and political discourse:

1. Mastering natural phenomena is the philosophical ideal of modern science and the direct manipulation and synthesis of natural objects represents the highest stage of experimental and theoretical progress. "Thus the will to dominate nature leads to a perspective of interpreting and manipulating nature, in which nature itself represents the essence of availability." (Kaulbach 1990) But as we have seen, the controllability under restricted conditions, which is an indispensable requirement for directed manipulations is tightly linked to uncertainty, once the altered objects leave the laboratory and enter the complexity and contingency of the "real" world. Many examples are known, where this newly created uncertainty was transformed into risk; hypothetical scenarios of unknown probability became real and had catastrophical consequences. Today's ecological problems, being caused at least partially by science based technologies and their products, demonstrate that the concept of domination over nature starts to brittle in the very moment, when mastery of natural phenomena seems to become perfect by the manipulation of life itself. What can be learned

from this is that decisions about specific theoretical and methodological approaches to natural phenomena are not only decisions about new ways to understand and to manipulate natural phenomena, but also about the way we constitute our theoretical and practical relationship to nature. And they are not only decisions about future benefits, but also about future risks.

The first conclusion which can be drawn from this is that the choice of a specific scientific perspective, that means the decision about what is included and what is excluded from our analysis of natural phenomena, is not only a scientific but also an ethical issue, and therefore must be questioned with regard to long term consequences for nature and society. This needs to be realized if we want to depict the relationship between science and morality, because it shows that value questions do not appear only in the course of application, but already in the course of generation of modern biological knowledge.

2. Regional or global health or environmental problems are in general a result of multiple, interwoven factors and/or developments. When scientific experts analyse such situations, they tend to define them according to well-known scientific patterns, so that the application for example of specific methods like genetic engineering appears essential. The perception and structuring of problems therefore is quite often already shaped by professional and personal points of view, methodological skills and political and social preferences. But it is not only that the presentation of an issue can be biased by professional and individual preferences. Moreover, one should be aware that due to their complex structure, which does not contain only scientific, but also social and political elements, real world phenomena cannot be adequately described by the terminology of natural sciences only.

Since it is known that different people - starting from various lifes and professional or personal experiences - have different perceptions of a situation, the analysis of a problem and the presentation of its results should include as many perspectives as possible. Such a pluralistic presentation is required as a necessary step prior to decision making, in order to ensure that the important aspects of a problem are recognized, and that possible consequences of a specific strategy are evaluated from a plurality of positions. An exclusive reliance on specialists from for example the natural sciences is at risk of leading into a one-eyed perception of the problems themselves. On these grounds, it is one of the most crucial points for policy related research projects, not to concentrate primarily on specific science based technologies as means to deal with problems, but to analyse the problems they are supposedly designed for themselves.

The second conclusion which can therefore be drawn is that the analysis and description of problems which need to be worked on or solved should not be left to one professional group only. Instead, one should try to get as many different portrayals of such a situation as possible, including those of citizens which are or will possibly be affected not only by the benefits, but also by the risks of the strategies which are proposed.

3. Technologies like genetic engineering and its products may have a longterm and irreversible impact on society and the environment. Due to public alertness it is not easy, to legitimate uncertain and risky experiments in and with the environment for the goals of scientific research alone. Therefore, such experiments are often declared to be the application of reliable knowledge in the pursuit of goals, the benefits or necessity of which it is hard to doubt. In connection with the deliberate release of genetically manipulated organisms the main goals named are: contributions for the solution of the problem of hunger worldwide, and the problem of environmental pollution.

But as we know, such experiments in the environment are not only about solutions for important global problems. Rather, they are also about specific interests of scientific research and the broadening of the sphere of activity and influence of science. They may also have a pathfinder role for a new technology, which may, after all not be the only, the most effective, or the safest one among those, which could be used. Taking this into consideration, we should remind ourselves that we are not inevitably bound to one specific technology. Therefore, it seems not only much wiser, but also much more fruitful, to consider and to develop different strategies for approaching the problems in question, since overall risks as well as benefits of a certain procedure cannot be evaluated *per se*, but only in comparison to other ways by which the same goals can be pursued. Another advantage of such an approach is that steps which need to be taken can be designed according to the structure and the requirements of the problem, and not according to those of a specific technology. They most likely involve a variety of different social and political means as well as scientific and technological instruments

Such an approach differs fundamentally from the one which was taken in 1985 by the inquiry commission on "Chances and Risks of Genetic Engineering" of the German Bundestag (Deutscher Bundestag 1987). The commission did practically not consider the argument that it may not primarily be a new technology which is needed for dealing with for example world nutritional problems, but instead or moreover political and economic strategies which are suited to ensure a more just and equal distribution of natural resources. It did not take into account alternative approaches to the problems to be worked on, although the methodology for doing this was already developed by the inquiry commission on "Future Nuclear Energy" in the late seventies. The work of this commission demonstrated that energy supply could be ensured under two different conditions of economic development, both with and without nuclear energy. (Deutscher Bundestag 1981) But although this work is something like a didactic piece of parliamentary work, unfortunately only parts of the results reached political practice.

This may primarily be due to the fact that it is extremely hard to change basic structures of decision making and politically and economically forced forward lines of technological development - especially against the ones, who are in power. Also, many people from the political or industrial establishment may not be willing to abandon basic believes and move into directions, which contradict or seem to contradict their views, although they may agree in specific measures which are proposed to be taken in order to avoid specific risks. It also could be the other way around: whereas it seems to be not very difficult to agree on basic values, the strategies to realize them could be highly controversial. But despite of such difficulties it should be kept in mind that science- or technology-centred strategies are not the only and probably not the best ones suited to work on environmental, agricultural, medical or social problems. Starting from different perspectives with respect to the perception and exposition on those problems, alternative pathways can be designed, and the existence of different possibilities of future developments can be demonstrated. This poses a challenging task to interested citizens as well as to many disciplines from the social and natural sciences, and other professions. Such an approach, which equally considers different strategies for problem solving, could be an important contribution to develop some new methods of risk assessment and to overcome some of the dead ends of the existing ones. With respect to rational decision-making it should therefore be avoided to concentrate on isolated technologies only and tried instead - and this is my third conclusion, to work out different pathways for further developments and compare the risks and benefits ascribed to them.

4. Thus there are at least three levels where the plurality of positions, perspectives and approaches is endangered by the primacy which was acquired by normal science and which has been given to solely scientific and technological strategies of problem solving: first on the level of competing epistemic positions, second on the level of competing social perspectives, and third on the level of competing political strategies of development.

Apart from economical arguments, one of the most important arguments for the primacy which is given to science based technologies is based on the perception that science is the most rational way to approach complex questions and problems which are supposedly brought about by natural causes. But we have seen on the one side that the problem of hunger worldwide can be perceived and described differently: not only as a problem of plants susceptible to pests or pesticides, which need to be manipulated by genetic engineering techniques, but instead as a problem of insatisfying agricultural techniques or of unequal distribution of food or means to buy it, just to name a few.

On the other side, we have learned that the objectivity and rationality of science are essentially based on a very small foundation, which is defined by the exclusion of real world complexity and contingency. The clearer these presumptions become by the historical, social and philosophical analysis of science, the more the claim of universality of scientific knowledge based on experiments under restricted conditions becomes questionable - at least for problems, which are situated outside the laboratory. Therefore, the scope of such knowledge has to be redefined. But this may be a tremendous chance to work out strategies, which do not follow the rationale of the laboratory only, but also include different epistemic, social and political perspectives and alternatives.

References

Bonß, W.; Hohlfeld, R.; Kollek, R. (1990)
Risiko und Kontext. Zum Umgang mit den Risiken der Gentechnologie. *Diskussionspapier Hamburger Institut für Sozialforschung* 5/90, 1-57

Bonß, W.; Hohlfeld, R.; Kollek, R. (to be published 1992)
Risiko und Kontext. Zur Unsicherheit in der Gentechnologie. In: Bechmann, Rammert (eds.), *Jahrbuch Technik und Gesellschaft*, vol 1991/1992. Campus, Frankfurt, pp 1-37

Deutscher Bundestag (1981)
Bericht der Enquete-Kommission "Zukünftige (Kern-)Energiepolitik". In: Deutscher Bundestag (eds.), *Zur Sache. Themen parlamentarischer Beratung,* vol 1/80., Bonn

Deutscher Bundestag (1987)
Bericht der Enquete-Kommission "Chancen und Risiken der Gentechnologie". In: Deutscher Bundestag (eds.), *Zur Sache. Themen parlamentarischer Beratung*, vol 1/87., Bonn

Falk, R. (1984)
The gene in search of an identity. *Human Genetics* 68:195-204

Häfele, W. (1974)
Hypotheticality and the New Challenges: The Pathfinder Role of Nuclear Energy. *Minerva* 12:303-322

Kaulbach, F. (1990)
Philosophie des Perspektivismus. Mohr, Tübingen

Kollek, R. (1988)
Gentechnologie und biologische Risiken - Stand der Diskussion nach dem Bericht der Enquetekommission "Chancen und Risiken der Gentechnologie". *WSI Mitteilungen* 2:105-116

Krohn, W.; Weyer, J. (1989)
Gesellschaft als Labor. *Soziale Welt* 40:347-359

Kuhn, T. S. (1967)
Die Struktur wissenschaftlicher Revolutionen. Suhrkamp, Frankfurt

Levins, R.; Lewontin, R. (1985)
The Dialectical Biologist. Harvard University Press, Cambridge

Nagl, W. (to be published 1992)
Grenzen unseres Wissens am Beispiel der Evolutionstheorie. Ethik und Sozialwissenschaften

Polanyi, M. (1968)
Life's irreducible structure. *Science* 160:1308-1312

Rayner, S.; Cantor, R. (1987)
How Fair is Safe Enough? The Cultural Approach to Societal Technology Choice. *Risk Analysis* 7:3-9

Sattler, R. (1986)
Biophilosophy. Analytic and holistic perspectives. Springer, Berlin, Heidelberg
Sturtevant, A. (1925)
The effects of unequal crossing over at the bar locus in drosophila. *Genetics* 10:117
Weinberg, A. M. (1972)
Science and Trans-Science. *Minerva* 10:209-222

3 PROBABILISTIC UNCERTAINTY AND TECHNOLOGICAL RISKS

Kristin S. Shrader-Frechette
University of South Florida

3.1 Introduction and overview

New applications of science and technology are typically saddled with high levels of uncertainty. Whether one is dealing with genetically engineered organisms, hazardous chemicals, or energy facilities, new applications of science and technology have not withstood the test of time. Safe periods of operating experience have not been established because, by definition, the applications are new, and many of their risks are unknown.

Fear is the classical public response to new and potentially dangerous applications of science and technology. Public fear, in such cases, inevitably raises the question of the rationality of lay responses to situations of scientific uncertainty. According to some technological risk assessors, public fear of new science and technology has much in common with our ancestors' fear of alleged witches. During the fourteenth through the sixteenth centuries, courts executed over a million "witches." According to risk assessor Alvin Weinberg, witch hunts subsided only after the Inquisitor of Spain convened a group of savants who proclaimed that there was no proof that "witches" caused misfortunes. Our current "environmental hypochondria," says Weinberg, is like the hysteria that drove witch hunts, and only savants realize that there is no proof that "environmental insult" causes "real health problems." Weinberg concludes that technological risk assessors need a new Inquisitor who is able to bring the public to its "senses."[1]

Like Weinberg, many scientists, risk analysts, and government policymakers have not dealt kindly with the public's distrust of high technologies and industrial toxins. A well-known energy spokesperson has condemned laypersons/-environmentalists as victims of "pathological fear" and "near-clinical paranoia."[2] Others have said that if the public only understood that catastrophic accidents were extremely unlikely, then they would not fear certain scientific and industrial activities.[3] Many technological risk assessors see themselves as the experts, the Inquisitors, who ought to bring ignorant and fearful laypersons to their senses.

Many risk assessors believe that laypersons will "come to their senses," in evaluating environmental risks, if they can learn to base their risk aversion on accident probabilities calculated by experts, rather than on their feelings. In other words, many assessors subscribe to "the Probabilistic Explanation."

3.2 The Probabilistic explanation

The "Probabilistic Explanation" is the belief that, for any rational and informed person, there is a linear relationship between a risk, defined as an actual probability of fatality (associated with a particular technological activity) and the value of avoiding the risk posed by that technology.[4] Following this strategy, many hazard assessors "explain" a societal aversion to certain low-probability technological risks by alleging that the public does not know the accident probabilities in question. They maintain that, given knowledge of the actual likelihood of death, rational persons always are more averse to high-probability risks than to low-probability ones.

In thus subscribing to the "Probabilistic Explanation," risk analysts likely err, in part, because the restriction of risk to "probability of fatality" is highly questionable. There are obviously many other cost burdens, e.g., "decreasing the GNP by a given amount," whose probability also determines the value of avoiding a given risk. Another problem is that the value of avoiding a given risk is often a function of the benefits to be gained from it, or whether it is distributed inequitably.[5] In fact, if Fischhoff and other assessors who employ psychometric surveys are correct, then risk acceptability is more closely correlated with equity than any other factors.[6]

Catastrophic potential and the fact that low- probability/high consequence situations are often the product of societally imposed (as opposed to privately chosen) risks may also explain risk aversion. There is evidence that the psychological trauma (feelings of impotence, depression, rage) associated with the imposition of a *public* hazard is greater than that associated with the choice of a *private* risk of the same probability. One author even suggests that widespread despair and an increasing suicide rate may be attributable to the hazards and fatalities caused by "industrial cannibalism."[7] If so, then there may be good reason why society's risk aversion is not proportional to probability of fatality. Moreover, although according to utility theory, a high-probability/low-consequence event (10,000 accidents, each killing 1 person) and a low- probability/high-consequence situation (1 accident's killing 10,000 persons) may have the same expected value, reasonable persons are typically more averse to the low-probability/high-consequence situation. One explanation may be that the high-consequence events, like catastrophic global warming, are often more difficult to quantify.[8]

If it is true that a risk's importance is not measured only by its probability, but is affected by other factors, then it makes sense for people to value the same level of safety differently in different settings. Even though this may be economically inefficient, it is neither inconsistent nor irrational. But if not, then why do risk assessors believe that societal aversion to allegedly low-probability risks is a consequence of false beliefs about the relevant probabilities?

3.2.1 Knowledge of probabilities and societal aversion to risks

To support their claims, many assessors maintain that laypersons view low-probability nuclear accidents as quite likely. Starr and Whipple, for example, citing the work of Otway, Lawless, and Fischoff, *et al.*, argue that "the bulk of disagreement" over nuclear power is over different beliefs about accident probabilities. Likewise, Cohen criticizes the public as being "uninformed" about the real risk probability from hazardous waste and obsessed with regulating risks that are "trivial."[9] Other assessors believe that, "Unlike the natural catastrophes -- earthquakes...etc., -- society has not learned to place such hypothetical man-made events [like nuclear catastrophes] in an acceptable comparative perspective, particularly when they are poorly understood by the public."[10] Generalizing on the basis of the nuclear-power case, they criticize public concern with "imaginary large catastrophes"[11] and suggest that conflicts over technology arise "because of intuitive estimates of unreasonably high risk" and because the public is "emotional" in its risk evaluations.[12] This essay will show that all these claims are questionable.

Although Starr and Whipple assert that Otway substantiates their claim about lay misperception of nuclear-accident probability (see note 9), Otway's studies show that he believes that the real controversy is over *values* and over incompatible views of the *benefits* attributed to nuclear power, not over different beliefs about accident probabilities. For example, Otway says that, "in general the con [nuclear power] group...assign high importance to the risk items while the pro group view benefit-related attributes as most important."[13] Otway also claims that, although both pro and con groups "strongly believe that nuclear power is in the hands of big government or business, ... the pro group evaluates this attribute positively, the con group evaluates it negatively."[14] In fact, Otway says explicitly that his research confirms the existence of only three statistically significant differences between pro-nuclear and anti-nuclear persons, all of which concern "the benefits of nuclear power." The pro group, Otway says, "strongly believed" that nuclear power offers three benefits: an essential good for society, economical energy, and a higher quality of life. On these three points, "the con groups tended to be uncertain to somewhat negative."[15] Regarding probabilistic evaluation of nuclear power, Otway explicitly states: "There were no significant differences between the [pro-nuclear and anti-nuclear] groups on the eb [evaluation-belief] scores of any items related to risk."[16] This conclusion appears to be a flat contradiction of the claim that Otway's research supports the Probabilistic Explanation.

Lawless' work does not seem to support it either. In fact, Lawless never mentions that misperceived probabilities cause disagreements over environmental risk. He argues instead that conflict over technology is greatest where proof of harm is *uncertain*, not where there is incorrect public perception of *certain* hazards.[17]

In the case of the recent controversy over methylene chloride (dichloromethane, DCM), for example, the dispute was clearly a function of *uncertainty* in scientific knowledge, not a result of incorrect public perception of *certain* knowledge. DCM is a multipurpose solvent used in paint stripping, metal cleaning, foam blowing, electronics, chemical processing, and in certain aerosol propellant mixtures. Because of its many applications, citizens are exposed to DCM in the workplace, through use of consumer products, and from emissions to ambient air. The U.S. EPA in 1987 said that DCM was a probable human carcinogen, although industry groups disagreed. The EPA cited the fact that DCM was carcinogenic in mice (although not so in hamsters and rats), whereas industry cited the fact that the two studies of humans exposed to DCM in the workplace showed no significant increase in cancer deaths.[18]

At the heart of the controversy over DCM between industry and environmentalists is the fact that industry tended to use pharmacokinetic models. These permitted the calculation of internal doses of DCM through integrated information on administered dose, the physiological structure of the species involved, and the biochemical properties of DCM. As a result, industry (e.g., Dow Chemical) predicted an average annual probability of fatality, for lifetime exposure to DCM, of 3.7×10^{-8}. (This risk is below the level of those typically regulated by government.) Other scientists, however, predicted the same risk as 4.1×10^{-6} for lifetime exposure to DCM. (Two orders of magnitude higher, this latter risk is at the level typically requiring government regulation.) Assessors obtained the higher risk figure by means of more conventional (than the pharmacokinetic) models. These involved linear extrapolation of external DCM dose and an interspecies factor based on body surface area.[19]

Although numerous other risk cases also exemplify the same point, the DCM controversy shows clearly that the dispute was not caused because the public incorrectly perceived the real, or probabilistic, risk. Indeed, even now (1990) the "real" level of risk is unknown (and cannot be calculated) because the carcinogenic mechanism, in the animals in which it has occurred, is unknown, according to the EPA.[20] Hence scientific uncertainty, not faulty public knowledge, appears to be driving the conflict.

Lawless also argues that the nuclear controversy, in particular, has been caused not only by scientific uncertainty, but also by the lack of credibility of federal regulators and by apparent government failure to consider environmental values. He notes, in general, that disputes over risk arose (in more than 50 percent of the cases studied) because technologies were allowed to grow, despite evidence that they were beset with problems, and because they were used irresponsibly.[21] All this suggests that laypersons' risk aversion may be reasonable, rather than merely a product of their erroneous risk perceptions.

As with the Otway and Lawless studies, there is little clear evidence, contrary to the claims of Starr, Whipple, and others, that the work of Fischhoff, *et al.* supports the thesis that environmental controversy arises because of lay misperception of risk probabilities. Their research with the League of Women Voters indicates that the commercial nuclear risk, for example, was not perceived as worth the benefits accruing from it.[22] A number of risk assessors, however, appear to dismiss the importance of the question of whether the *nuclear* benefit is worth the risk (see endnote 51). Since these same assessors claim that risks and benefits are not evaluated independently, however, it is not clear how they can be so sure that the debate over nuclear safety is primarily over probabilities, rather than whether the benefit is worth the risk.

Fischhoff, *et al.*, specifically note that, on their surveys, the public (students and members of the League of Women Voters) judged nuclear power to have "the *lowest* fatality estimate" for the 30 activities studied, but the "*highest* perceived risk."[23] As a consequence, they conclude, "we can reject the idea that lay people wanted to equate risk with annual fatalities, but were inaccurate in doing so. Apparently, laypeople incorporate other considerations besides annual fatalities into their concept of risk."[24]

According to the Fischhoff study, the key consideration influencing judgments of high risk was not perceptions of high accident probability but the fact that certain technologies represent an unfamiliar (as opposed to a common) risk; an inequitably (as opposed to equitably) distributed risk; and a hazard with severe consequences in the unlikely event that an accident were to occur.[25] In fact, the assessors found that perceived risk could be predicted almost completely accurately solely on the basis of the single variable, "severity of consequences," even though the probability of those consequences' occurring was quite small and was perceived as quite small.[26] If this is true, then the suggestion that environmental controversy is fueled primarily by incorrect *probability* estimates of laypersons is less helpful than the suggestion that, in cases of high-magnitude events, it is the possible *consequences* that are important to societal evaluation. Wilson proposes that *N* lives lost simultaneously in a catastrophic accident should be assessed as a loss of N^2 lives. He argues that the risk-conversion factor for catastrophic accidents should be exponential.[27]

Admittedly, the studies by Fischhoff and others show only that, contrary to the assertions of other assessors, controversy over hazardous technologies very likely arises because of the value placed on consequences, not because of overestimated risk probabilities. The studies do not show that consequences ought to be valued in this way. Moreover, the Atomic Energy Comission (AEC), the Nuclear Regulatory Commission (NRC), and the courts generally have not attributed much importance to consequences. Courts "have consistently taken the position that probabilities are determinative of risk, regardless of potential consequences."[28] Nuclear risk assessments have also consistently adopted the nuisance rule that probabilities alone determine risk, probably because of society's interest in technological development. Historically, the rule has owed its inspiration to the reluctance of the nineteenth-century courts to allow the traditionally restrictive law of nuisance to hinder economic progress.[29]

3.2.2 Do probabilities, alone, adequately define 'risk'?

There are, however, a number of reasons for arguing that, in certain cases, risk consequences are more important than the accident probabilities. For one thing, greater social disruption arises from one massive accident, as compared to many single- fatality accidents killing the same number of people. The law of torts also recognizes the heightened importance of high-consequence events, apart from their probability of occurrence. In fact, for the rule of strict liability, risk is based almost totally on grave potential consequences, regardless of the associated probability.[30] Part of the justification for this judicial emphasis on accident consequences is apparently the fact that the parties involved in litigation over catastrophic accidents -- viz., the injured persons and those liable for the injury -- are not equal in bargaining power.[31] The representative of some technological or industrial interest usually has more clout than the person damaged by it. Moreover, a person is more deserving of compensation according to strict liability when she is victimized by an impact that she did not voluntarily accept or help to create. And if so, then societal risk evaluation of potentially catastrophic technologies ought to focus on the accident consequences, as well as on their probabilities.

This point is clear if one considers a rational response to the invitation to play Russian Roulette. Suppose the probability that a bullet is in a chamber when the trigger is pulled is one in 17,000 -- the same likelihood, per reactor-year, as a nuclear core melt. Even with such a small probability, a person could still be rational in her refusal to play the game. She could even maintain that the probability in question is irrelevant. Any probability of fatality might be too high, if the benefits deriving from taking the risk were not great enough. And if so, then probabilities might not be as important, in environmental risk evaluation, as proponents of the Probabilistic Explanation suggest. As one expert expressed it, current debate over whether a given technology has a particular risk probability is a spurious issue. "Risk assessors tend to choose methods and data that support the position to which they are already committed."[32] But if so, then debate over environmental risks is likely to be over many factors other than probability.

3.3 Risk controversy and beliefs about probabilities

The claim that probabilities are central to risk evaluation, and that "the bulk of disagreement" over environmental hazards has been caused by "intuitive estimates of unreasonably high risk," also errs in ignoring reasonable disagreement over risk probabilities. Risk assessment has been repeatedly criticized as an "arcane expert process" overly dependent on probability estimates of assessors.[33] Many risk assessors appear to believe that it is "perfectly valid to base public policy on expert estimates and data," but that, once a risk expert has spoken, any disagreement is unreasonable and intuitive.[34] Such a notion is doubly questionable.

It is in part implausible because it presupposes a far more objective picture of probabilistic risk data than is now available. Even the authors of the most complete hazard analysis ever accomplished, WASH-1400, cautioned that their probability estimates were deficient, unprovable, possibly incomplete, assumption-laden and saddled with "an appreciable uncertainty." They said that "the present state of knowledge probably will not permit a complete analysis of low-probability accidents in nuclear plants with the precision that would be desirable.[35] More generally and more recently, risk assessors have pointed out that "uncertainties of six orders of magnitude are not unusual" in any probabilistic risk analysis.[36] In the face of such *caveats*, alleged certitude about which risk probabilities are correct, and which are incorrect, may be doubtful. Often the scientific mechanisms causing a hazard are unknown, as in the case of methylene chloride.[37] Moreover, *accident probability* often cannot be determined on the basis of observed *accident frequency*.[38] On the one hand, very low values of an accident probability per LNG trip, or per reactor-year, for example, are consistent with an assumed record of zero accidents in 800 voyages, or zero core melts in 17,000 reactor years. On the other hand, an annual accident probability as high as 1 in 100 or 1 in 200 would still be consistent with the current LNG accident- frequency record, just as a yearly probability as high as 1 in 2000 would be consistent with the existing nuclear-accident record. Even though an accident record may be consistent with very low risk probability values, this frequency alone "does not prove that the values are low."[39]

Proponents of the Probabilistic Explanation also err, in emphasizing risk probabilities, because they are unable to account for the reasonable controversy, among Nobel Prize winners, the American Physical Society (APS), the Environmental Protection Agency (EPA), the Nuclear Regulatory Commission (NRC), and the American Nuclear Society (ANS), over various risk probabilities.[40] Reputable assessors affirm that many of the most serious environmental risks, e.g., global warming from burning of fossil fuels, are "highly resistant to quantification."[41] Moreover, there are a number of difficulties which make nuclear probabilities, for example, especially resistant to accurate estimation. Compound events, sequential component failures, sabotage or human error, and weapons' proliferation are not amenable to quantification.[42] Rasmussen himself computed the probability of having a Three-Mile-Island- type accident as anywhere from 1 in 250 to 1 in 25,000 reactor- years.[43] All this suggests that certain accidents are not really "impossible," because many low probabilities are not believable. For example, the probability for a royal flush is 1 in 464,000. Yet, in a card game the probability is actually much higher, since the probability of cheating is likely to be as high as 1 in 10,000. Likewise, although the probability of a given environmental or technological accident may be only very slight, the higher probability of sabotage or terrorism is likely to increase this number by several orders of magnitude. This means that real risks are likely to include so-called "outrageous events" or "rogue events" which are difficult to handle in probabilistic risk assessment.[44] Indeed, human error causes a majority of most industrial, marine, and transportation accidents.[45]

In claiming that the public overestimates many risk probabilities, like those for nuclear accidents, many assessors assume that, in some of the most controversial, untested, and potentially catastrophic areas of technology, it is possible to judge clearly when a risk probability is accurate and when it is not. This is an appeal to authority, an appeal which (given the history of science) simply does not hold up.

Risk assessors' emphasis on the importance of probability estimates is especially vulnerable when one recalls that the characteristics hypothesized by various authors to influence judgments of perceived and aceptable risks are highly inter- correlated; involuntary hazards, for example, "tend also to be inequitable and catastrophic."[46] This means that it is especially difficult to determine whether or not society's expressed concern about involuntary risks, for example, is merely an artifact of the high correlations between involuntariness and other undesirable risk characteristics. There are numerous allegedly causal explanations, all consistent with the same "observed" phenomena. Kasper made an analogous observation:

> Even the best of epidemiological studies is confounded by the myriad explanations for low-level neurobehavioral effects;the same effects attributed to leadmay be caused by exposure to low levels of many other trace metals, and indeed by exposure to the pace and stress ofurban life itself. The resultis that careful studies yield not proof but only suggestions.[47]

Or as another risk assessor put it, "multiple models, having quite different implications at low doses, may all adequately 'fit' the observed dose-response data."[48]

Precisely because their hypothesis about misperceived probabilities is consistent with other explanations, proponents of the Probabilistic Explanation are not warranted in singling out the public's alleged misperceived probabilities as the *cause* of its high aversion to societal risks. Rather, the distinction between expert/objective, versus lay/subjective, determination of environmental risks will not hold up. Because of problems of actual risk *calculation* (prior to any alleged evaluation), many hazard estimates are merely the intuitive *guesses* of individuals. Authors of a recent study done at the Stanford Research Institute admitted, for example, that analytical techniques could not handle probability estimates for certain human-caused events. They concluded: "We must rely on expert judgment, quantified, using subjective probabilities."[49] Likewise, the loss of the astronauts in the Challenger disaster, as well as the death of three astronauts on the ground at Cape Kennedy demonstrated that even the best systems-analytic approaches cannot anticipate every possibility. In fact, one of the most famous nuclear risk probabilities, widely touted as "objective," is highly value- laden. This is the reactor-year probability of a core melt in a nuclear plant, 1 in 17,000. As defended in WASH-1400, this probability is notoriously laden with value judgments about the effectiveness of evacuation in the face of catastrophe, the probability of weather stability, and the Gaussian Plume rise of radioactivity.[50] The problem, however, is *not* that such "objective" probabilities (as given by experts) are value-laden, but that they are apparently not recognized as such by proponents of the Probabilistic Explanation.

3.4 Lessons learned from experts' Claims about societal risks

The tendency of proponents of the Probabilistic Explanation, to overemphasize the importance of risk probabilities and to condemn the public's alleged "misperceptions" of societal risks, reveals an important flaw in contemporary environmental risk analysis. Assessors presume that, if there is a public preference for a risk whose probability-of-fatality is statistically higher than that of an alternative, then this preference is a result of misperceived probabilities, not a legitimate value system. Their failure to recognize the value components of allegedly objective probability estimates goes hand-in-hand with assessors' tendencies to define *ethical* and *political* issues as merely *technical* ones, as the naive positivists are prone to do.[51] They assume, incorrectly, that agreement about technical matters is sufficient for resolving normative disputes.

Apparently they make this assumption because they are afraid of damaging "the scientific pretenses of their work."[52] As a consequence, their emphasis on the importance of abstract, "objective" science helps both to disguise the often exploitative way in which technology is used, and to condone a passive acceptance of the *status quo*. It allows assessors to dismiss as irrational or unscientific, as Okrent, Starr, Whipple, Maxey, Cohen, Lee, and others have done, any attempts to challenge our contemporary ethical or political values.[53] But as Dickson has argued, "the use of supposedly objective models of...social behavior serves to legitimate the imposition of social policy."[54] This is because the real risk concerns of laypersons can then be dismissed as subjective. As one critic put it, this is "like playing Monopoly with the Mafia: they always start the game owning Boardwalk."[55]

3.5 Conclusion

Because experts often define risk only in terms of probability of fatality and consequently neglect ethical and political concerns, they fail to attack an essential problem of risk evaluation: how to make the decision process more *democratic*. As Thomas Jefferson warned, the only safe locus of societal power is in the people. He wrote: "I know of no safe depositor of the ultimate powers of the society but the people themselves; and if we think them not enlightened enough to exercise their control with a wholesome discretion, the remedy is not to take it from them, but to inform their discretion."[56]

Notes

1. A. Weinberg, "Risk Assessment, Regulation, and the Limits," in *Phenotypic Variation in Populations*, ed. A. Woodhead, M. Bender, and R. Leonard (New York: Plenum, 1988), pp. 121-128; hereafter cited as: Weinberg, Risk 1988, in Woodhead, Variation 1988.

2. M. Maxey, "Managing Low-Level Radioactive Wastes, in *Low- Level Radioactive Waste Management* (Williamsburg, Virginia: Health Physics Society, 1979), pp. 400-409.

3. See, for example, B. Cohen, "Risk Analyses of Buried Wastes," in *The Risk of Environmental and Human Health Hazards*, ed. D. J. Paustenbach (New York: John Wiley, 1989), p. 575; hereafter cited as: Cohen, Risk 1989, in Paustenbach, RA 1989. See also C. Whipple, "Nonpessimistic Risk Assessment," in Paustenbach, RA 1989, pp. 1112-1113; hereafter cited as: Whipple, Risk 1989. Finally see B. Cohen and I. Lee, "A Catalog of Risks," *Health Physics* 36, no. 6 (1979): 707; W. Hafele, "Energy," in *Science, Technology, and the Human Prospect*, ed. C. Starr and P. Ritterbush (New York: Pergamon, 1979), p. 139; Starr, "Benefit- Cost Studies in Sociotechnical Systems," in *Perspectives on Benefit-Risk Decision Making*, ed. Committee on Public Engineering Policy (Washington, D.C.: National Academy of Engineering, 1972), p.26-27, hereafter cited as: BCS; and L. Lave, "Discussion, in *Symposium/Workshop...Risk Assessment and Governmental Decision Making*, ed. Mitre Corporation (McLean, Virginia: Mitre Corporation, 1979), p. 484, hereafter cited as: Symposium.

4. See, for example, Cohen, Risk 1989, p. 575, and Whipple, Risk 1989, pp. 1112-1113. See also K. S. Shrader-Frechette, "Economics, Risk-Cost Benefit Analysis, and the Linearity Assumption," in *PSA 1982*, ed. P. Asquith and T. Nickles (East Lansing, Michigan: Philosophy of Science Association, 1982), and K. S. Shrader-Frechette, *Risk and Rationality* (Berkeley: University of California Press, 1991), ch. 6.

5. L. Cox and P. Ricci, "Legal and Philosophical Aspects of Risk Analysis," in Paustenbach, RA 1989, pp. 1017-1046; hereafter cited as: Cox and Ricci, Legal 1989. See also W. Rowe, *An Anatomy of Risk* (New York: John Wiley and Sons, 1977), p. 926; hereafter cited as: Anatomy.

6. B. Fischhoff, P. Slovic, and S. Lichtenstein, "Facts and Fears," in *Societal Risk Assessment*, ed. R. Schwing and W. Albers (New York: Plenum, 1980), p. 207; hereafter cited as: FF and SRA. See also R. Kates and A. Weinberg, *et al.*, *Hazards: Technology and Fairness* (Washington, D.C.: National Academy Press, 1986), Part 2; hereafter cited as: Kates, Hazards 1986.

7. S. Samuels, "The Arrogance of Intellectual Power," in Woodhead, Variation 1988, pp. 113-120; hereafter cited as: Samuels, Power 1988. See also P. D. Pahner, "The Psychological Displacement of Anxiety: An Application to Nuclear Energy," in *Risk-Benefit Methodology and Application*, ed. D. Okrent (Los Angeles: UCLA School of Engineering and Applied Science, 1975), p. 575; hereafter cited as: RBMA.

8. P. Gleick and J. Holdren, "Assessing the Environmental Risks of Energy," in *American Journal of Public Health* 71, no. 9 (September 1981): 1046; hereafter cited as: Gleick and Holdren, Risk 1981. See also A. J. Van Horn and R. Wilson, "The Status of Risk-Benefit Analysis," discussion paper (Cambridge, Massachussetts: Harvard University Energy and Environmental Policy Center, 1976), p. 19.

9. B. Cohen, Risk 1989, pp. 575-575. See also C. Starr and Whipple, "Risks of Risk Decisions," *Science* 208, no. 4448 (June 6, 1980): 1116; hereafter cited as: Risks.

10. Starr, BCS, pp. 26-27.

11. Starr, BCS, pp. 29-27.

12. Cohen, Risk 1989, p. 575. Starr and Whipple, Risks, p. 1116; and Starr, *Current Issues in Energy* (New York: Pergamon Press, 1979), pp. 16-17.

13. H. Otway, "Risk Assessment and the Social Response to Nuclear Power," *Journal of the British Nuclear Engineering Society* 16, no. 4 (1977): 331; hereafter cited as: Response.

14. Otway, Response, p. 331.

15. Otway, Response, p. 332.

16. Otway, Response, p. 332.

17. E. Lawless, *Technology and Social Shock* (New Brunswick, N.J.: Rutgers University Press, 1977), pp. 497-498, 512; hereafter cited as: TSS.

18. J. W. Falco and R. Moraski, "Methods Used in the United States for the Assessment and Management of Health Risk Due to Chemicals," in *Risk Management of Chemicals in the Environment*, ed. H. M. Seip and A. B. Heiberg (New York: Plenum, 1989), pp. 37-60; hereafter cited as: Falco and Moraski, Methods 1989, in seip and Heiberg, Risk 1989.

19. R. Reitz, F. Smith, *et al.*, "Use of Physiological Pharmacokinetics in Cancer Risk Assessments," in *The Risk Assessment of Environmental and Human Health Hazards*, ed. D. J. Paustenbach, RA 1989, pp. 238-265, esp. p. 258; hereafter cited as: Reitz, Risk 1989.

20. Falco and Moraski, Methods 1989, p. 55.

21. Lawless, TSS, pp. 349-357, 434-435, 490.

22. See note 6 and B. Fischhoff, P. Slovic, S. Lichtenstein, S. Read and B. Combs, "How Safe is Safe Enough?" *Policy Sciences* 9, no. 2 (1978): 150; hereafter cited as: HSSE. See also Cox and Ricci, Legal 1989, pp. 1017-1046.

23. Fischhoff *et al.*, FF, p. 192.

24. Fischhoff *et al.*, FF, p. 192.

25. Fischhoff *et al.*, HSSE, pp. 140-142, and FF, p. 202. See also R. Kasper, "Perceptions of Risk and Their Effects on Decision Making," in Schwing and Albers, SRA, p. 75; hereafter cited as: Perceptions.

26. Fischhoff *et al.*, HSSE, pp. 148-149; and H. Green, "Cost- Benefit Assessment and the Law," *George Washington Law Review* 45, no. 5 (1977): 909-910. See also L. Clarke, *Acceptable Risk?* (Berkeley: University of California Press, 1989), pp. 178-182; hereafter cited as: Clarke, Risk 1989.

27. Fischhoff *et al.*, FF, p. 208; and Rowe, Anatomy, p. 290. See Cox and Ricci, Legal 1989, pp. 1036-1037.

28. J. Yellin, "Judicial Review and Nuclear Power," *George Washington Law Review* 45, no. 5 (1977): 992; hereafter cited as: JRNP.

29. Yellin, JRNP, p. 987. See Whipple, Risk 1989, p. 1111.

30. Yellin, JRNP, pp. 983-984.

31. Yellin, JRNP, pp. 987-988. See P. Huber, "The Bhopalization of American Tort Law," in Kates, Hazards 1986, pp. 89-110; hereafter cited as: Huber, Tort 1986.

32. Clarke, Risk 1989, p. 181.

33. R.N. Andrews, "Environmental Impact Assessment and Risk Assessment," in *Environmental Impact Assessment*, ed. Peter Wathern (London: Unwin Hyman, 1988), pp. 85-97; hereafter cited as: Andrews, RA 1988.

34. Starr and Whipple, Risks, p. 1116.

35. NRC (U.S. Nuclear Regulatory Commission), *Reactor Safety Study: An Assessment of Accident Risks in U.S. Commercial Nuclear Power Plants*, WASH-1400 (Washington, D.C.: U.S. Government Printing Office, 1975), pp. 15, 40, 96-97, 224; hereafter cited as: NRC.

36. Cox and Ricci, Legal 1989, p. 1027.

37. D. Okrent and C. Whipple, *Approach to Societal Risk Acceptance Criteria and Risk Management*, PB-271264 (Washington, D.C.: U.S. Department of Commerce, 1975), p. 10. For methylene chloride risk information, see Falco and Moraski, Methods 1989, and Reitz, Risk 1989.

38. W. Lowrance, "The Nature of Risk," in Schwing and Albers, SRA, p. 6; hereafter cited as: NR. See Rowe, Anatomy, p. 264.

39. W. Fairley, "Criteria for Evaluating the 'Small' Probability," in Okrent, RBMA, p. 425; hereafter cited as: Criteria.

40. See notes 18 and 19. See also NRC, Appendix XI, 2-1--2-14; W. Hafele, "Benefit-Risk Tradeoffs in Nuclear Power Generation," in *Energy and the Environment: A Risk-Benefit Approach*, ed. H. Ashley, R. Ashley, R. Rudman, and C. Whipple (New York: Pergamon, 1976), pp. 159-169; Lieberman, GH, pp. 250-255. See Whipple, Risk 1989, and Cohen, Risk 1989.

41. Gleick and Holdren, Risks 1981, p. 1046.

42. Gleick and Holdren, Risks 1981, pp. 1046, 1049. R. Zeckhauser, "Procedures for Valuing Lives," *Public Policy* 23, no. 4 (1975): 445; Committee on Public Engineering Policy, *Perspectives on Benefit-Risk Decision Making* (Washington, D.C.: National Academy of Engineering, 1972), p. 10. See also Lowrance, NR, p. 11.

43. N. C. Rasmussen, "Methods of Hazard Analysis and Nuclear Safety Engineering," in *The Three Mile Island Nuclear Accident*, ed. T. Moss and D. Sill (New York: New York Academy of Sciences, 1981), p. 29.

44. Fairley, Criteria, p. 406-407.

45. L. Philipson, "Panel," in Mitre Corporation, Symposium, p. 246.

46. Fischhoff *et al.*, HSSE, pp. 144-149, 215. See Cox and Ricci, Legal 1989.

47. Kasper, Perceptions, p. 73. D. Cleverly, *et al.*, *Municipal Waste Combustion Study* (New York: Taylor and Francis, 1989), pp. A-10, 2-6, 2-12, and 3-18, for examples of these uncertainties.

48. L. Cox and P. Ricci, "Risk, Uncertainty, and Causation," in Paustenbach, RA 1989, p. 151; See also L. Maxim, "Problems Associated with the Use of Conservative Assumptions in Exposure and Risk Analysis," in Paustenbach, RA 1989, pp. 526 ff.

49. A. Lovins, "Cost-Risk-Benefit Assessment in Energy Policy," *George Washington Law Review* 45, no. 5 (1977): 926. Similar problems are expressed, for example, by J. Harkins, E. Scott, and W. Walsh, "A Legal Viewpoint," in Woodhead, Variation 1988, pp. 218 ff.; see also R. Cortesi, "Variation in Individual Response," in Woodhead, Variation 1988, pp. 288-289.

50. NRC, pp. 108-109, 118, 186, 239, 245-246.

51. See Cohen, Risk 1989, p. 575, and Whipple, Risk 1989, pp. 1111-1113. See also K. S. Shrader-Frechette, *Risk and Rationality* (Berkeley: University of California Press, 1991), ch. 3.

52. H. Stretton, *Capitalism, Socialism, and the Environment* (Cambridge, England: Cambridge University Press, 1976), p. 51; hereafter cited as: CSE. See Andrews, RA 1988, pp. 85-97.

53. D. Dickson, *The Politics of Alternative Technology* (New York: Universe Books, 1975), p. 189; hereafter cited as: PAT. See also See also K. S. Shrader-Frechette, *Risk and Rationality* (Berkeley: University of California Press, 1991), chs. 2 and 5.

54. Dickson, PAT, and Stretton, CSE.

55. Samuels, Power 1988, p. 118.

56. Cited in D. Bazelon, "Risk and Responsibility," *Science* 205, no. 4403 (1979): 277-280.

PART II

Concepts of Science for Policy

4 A NEW BRANCH OF SCIENCE, INC.[1]

Helga Nowotny
University of Vienna, Austria

4.1 The separation of science from public policy

Those responsible for science policy occasionally run the risk that a piece of unanticipated reality may be lurking behind the metaphorical imagery they have constructed in order to accommodate a broad spectrum of different ideas. The conventional link between science and public policy is to think in terms of public policy for science -a long-standing concern among a small circle of experts drawn from the natural sciences, the policy sciences, and politicians as to how to find optimal ways of funding research and of guiding the innovative process of scientific-technological development. Yet, the converse combination is also possible, namely to think of science for public policy. This has, as I will try to show, both an obvious ring of familiarity, asking us to restate and perhaps clarify the directive mission contained in the pronoun, but at the same time a more provocative meaning inviting us to overcome the *de facto* separation of science from public policy.

Let me first consider the obvious meaning: science for public policy as the outgrowth of the oldest social mission of science -for the public good. Ever since the inception of modern science in seventeenth-century England, with the incisive formulations of Francis Bacon, scientists and technologists have conceived their activities in terms of noble aspirations. By linking their work to an increase in welfare -first of their own nations, but ultimately of the entire human race- they sought to reduce suffering due to the lack of means, to satisfy material wants, and to alleviate degrading labour. The collective purpose of science conceived in these broad terms has hardly changed. In the latter part of the twentieth century the common good is still on the public agenda and policies are still directed towards tangible results. As Harvey Brooks has reminded us, the standard list of fundamental human needs to which science and technology are expected to contribute is still remarkably unchanged: food and energy supply, health needs, transportation, shelter, personal security. Later additions seem to be the remaining items: a cleaner environment and a social system which, in the words of Harvey Brooks, facilitates rapid adaptive change while restraining the possibility of violent conflict.[2]

Such additions to the standard list already signal the shift from the tangible results of science and technology, from their expected direct contributions to economic growth and welfare, to the more intangible, indirect, and mediated ones. Today, science for public policy can no longer concentrate on accelerating the rate of innovation as an aim in itself. Rather, it has increasingly become preoccupied in dealing with the unwanted and unintended effects of its direct contributions. The quest for a cleaner, safer environment is a case in point. The secondary and tertiary effects, the as-yet unknown consequences, of our interaction with the environment have become the source of our main concerns. There is an equal quest for a social system that would facilitate adaptive change and yet not be overturned by it. The expected contributions of science for public policy have shifted from the operational to the symbolic realm. Utilizing its cognitive capacities, putting knowledge as the most precious resource science has to offer at the disposal of policymakers thought to be in desperate need of it, scientific knowledge and information has become the key for managing a future whose existence is threatened by the intervention made in the past. Science, so far, holds an absolute monopoly on this kind of knowledge and, as other previous monopoly holders, it has to maintain its claims by guarding its institutional boundaries, in this case, its autonomy in the production of knowledge. This is one reason why the dividing line, separating scientific facts from values, ordinary everyday knowledge from scientific knowledge, scientific expertise from lay participation, and science from politics, is so entrenched. What science had to offer -according to its own definition of its social mission- was advice: advice held to be clean from political considerations, free from values and mere opinions, from interests and control over its later applications. Science was disinterested and neutral, committed solely to its own impartial and context-independent conception of Truth. This, at least, was the ideal.

But is such a formula sufficient? Is this what science for public policy is all about, when the pressure of taking action mounts is areas of genuine scientific uncertainty, and when the roles of what once were thought to be "hard" scientific facts amid "soft" human decision-making procedures, as Jerome Ravetz has pointed out[3], are becoming reversed and we now are confronted with the necessity for making "hard" decisions in the face of "soft" scientific evidence? While science for public policy is firmly ingrained in the social mission of science, both in the sense of tangible, instrumental results and the more intangible resource of providing information for guiding the policy processes, the lines separating science *from* public policy are also sharply drawn. Harvey Brooks states this very clearly: science and technology, he writes, cannot provide a solution by themselves. They can only generate the conditions in which a society can develop a solution[4].

But does the policy process really live up to the expectations put into it? Who does the translation from one field to the other in the first case and what happens (as invariably it does happen) if scientific findings get transformed, distorted, subject to political bargaining in the translation process? Is it really true that science "only" creates the conditions in which society can develop a solution? Are not both science and the evolution of an institutional societal framework geared towards the production of certain types of solutions, linked to each other through a common historical ancestry? Are not both, as Max Weber suggested a long time ago, embedded in the process of ongoing rationalization that happened to be both a precondition and the most important consequence for capitalism to evolve, bent on achieving a high degree of predictability and calculability, of efficiency in the domain of nature as well as within the social and economic order? While the spillover effects of the scientification of everyday life, including political institutions, has been enormous, one ought not to lose sight of the tremendous changes that science, its organization, and the concept of science have undergone in this very same process.

Thus, the innocuous looking line that restates the obvious -that science is for public policy- while at the same time separating science from public policy -by claiming that it only creates the conditions for society to develop solutions- open up a dilemma which is becoming more acute under the pressure for new solutions on the part of science for public policy.

4.2 How rational is the policy process?

The impact of the process of rationalization has been uneven: while the organization of scientific knowledge became the model of rational organization *per se*, the political process is generally viewed as lagging far behind. It is worthwhile to recall the great appeal that the scientific method once commanded as a way of settling disputes, and the futile hope that was expressed again and again, in scientific and political utopias alike, that it would be possible to arrive at similar rational procedures for solving conflicts in the political realm[5]. The dominant view of science for public policy shares some of these elements, since it rest on the implicit assumption of an underlying structural similarity of mutually converging rationalities. This assumption has been elaborated in two direction: one is the still dominant model of rational decision making that was devised especially by policy analysts, and the other one is the view that a great number of scientists hold about the nature of their input into the policy process.

This picture of a rational or, perhaps better, over-rationalizing model of the policy process has not failed to repeatedly attracts well-founded criticism. Majone, among others, has pointed to an underlying deeper commitment to a teleological, end-result conception of policy making and the reliance upon a number of fictional constructs which follow from the model[6]. In a thorough review, Aaron Wildavsky highlights the essential difference that exist between puzzles -to be solved once and for all- and (policy) problems that may be alleviated, eventually superseded, and finally redefined. He declares that the "rational paradigm" is simply mistaken. It fails to adapt to the ways in which decisions are actually made, where available answers determine the kinds of questions that are asked and objectives are never the products of the seat of rationality, but dependent upon available resources[7]. Others, like Peter House, have systematically questioned the assumptions by which policy analysis was supposed to be brought into the policy process, by comparing a number of actual cases with their analytical foundations[8]. In attempting to explain why policy-oriented research seems to have had little or no direct impact on policy making, Björn Wittrock has suggested that the mismatch between the supply and the use of policy-relevant social knowledge can be traced either to a highly rationalistic conception of the policy process -the "social engineering" model- or to an "enlightenment" model that assumes that social science research does not so much solve problems as provide an intellectual setting of concepts, orientations, and empirical generalizations. He argues in favour of a third model -a dispositional one- a conception of knowledge utilization: the process is neither arbitrary and haphazard, nor entirely pre-programmed; important policy research must be there to be utilized and if conditions are propitious and important actors available, its findings might well have an impact[9].

While some of these commentaries and criticisms pertain more to the utilization of social science knowledge, there is widespread recognition of the enduring and conflicting nature of public policy issues in general which have increasingly come to include environmental and technological issues[10]. In such an enlargened view of policy analysis, the question of the epistemological foundation is also receiving renewed attention. Thus, in a recent review of policy research and a rejoinder undertaken in defense of the policy sciences as science, one consistent theme of contention between the authors was that one of their models would follow an outdated positivistic conception of science, while the policymaking process should be viewed as resting on a much broader epistemological basis[11].

While some of this ongoing dismantling of the Received View can be interpreted as a necessary correction of the immature field of policy analysis, I think that the reasons lie somewhat deeper. The Received View has been adopted not only by its proponents -over-confident about rational problem solving and about the extension of methods and tools from one realm- that of military and industrial operations -to the much more complex and ambiguous arena of political and social issues- but also has its adherents among actual decision makers and scientists alike. It conformed to the Enlightened View that science and public policy were either slowly converging in their inherent rationalities or that public policy, in order to be receptive to scientific advice and improvement, had to come to resemble more closely what a scientific model of the policy process demanded it to be. This was a highly convenient way of thinking about science for public policy, as long as it remained the exclusive concern of a relatively small circle of public policy officials and scientists involved as advisors in certain policy arenas. It fitted into an institutional arrangement, moreover, that defined public policy as falling within the competence of a relatively closed administrative-scientific coalition.

Not surprisingly, the correlative view held by many scientists involved in the policy-making process as experts or advisors carries an equally strong faiths to what good public policy is all about. It is to be guided by scientific expertise which claims authority also over the definition of good government: one that admits to strong scientific guidance in how to conduct political affairs. There was a recent reanalysis of the testimony of some 130 expert witnesses who stated their views on the necessity and desirability of creating a US Congressional Office for Technology Assessment. Most of these witnesses were of the opinion that technology is to be equated with effective intelligence which they considered to function as a substitute for an otherwise failed sense of history, of logic and purpose in the unfolding of events[12].

Although expressed in a particular context and referring explicitly only to technological expertise, such views probably accurately reflect the confident attitude of a scientific-technological élite involved in the public policy process so long as their equally held belief in the impartiality of their expertise remained unchallenged. Good science and good public policy would meet as long as both would conform to the underlying assumption of the growing convergence in their rationality. The shock and disturbance which came with public contestation were accordingly great.

4.3 Science contested: science for whom?

As long as public trust in science and technology was still high and undisturbed, as long as it was a small circle of a scientific élite that functioned as advisors to governments and administrative officials, as long as the public image of science would reasonably cover what scientists themselves projected their activities to be for society -science for public policy was what good scientists did for a rational policy process[13]. The internal hierarchy of the status system was sufficiently strong to carry its weight in the public arena and the internal status system determined who a good scientist was. Looking back to the time before public contestation seems almost like looking back at a bygone age. Science and public policy have long since ceased to be bound by a relationship consisting simply of a few representatives of science and a few policy makers and officials. The public has intruded in public policy and is, so it seems, here to stay, even though it is not always easy to say who the public is. Most observers would most probably agree that a new set of political actors and new social movements have come to the fore in the stream of an altered public awareness of the impact of science and technology. They have done so, first by questioning what has been taken for granted so far, namely that science always works for better public policy; then by protesting that their concerns were not taken into account properly; and finally by claiming that science for public policy should be subject to participatory scrutiny like other inputs into the political process. Since it had become obvious that science and technology could sometimes have negative side effects and even potentially cause great harm, the assumption valid from the seventeenth century onwards that science would inevitably produce results for the public good has definitely come to its end.

Among the many repercussions that the protest phase had on the relationship between science and public policy, I want to single out those that illustrate the changes of the context in which science for public policy is defined today. This changed context reflects a new balance of the tensions inherent between science and public policy.

The first outcome is the undermining of the alleged rationality of the political process, which turned out to be far less rational than depicted by the champions of rational policy analysis. There was not one unitary decision maker but a multitude of conflicting parties. The political process showed itself to a certain degree receptive to protests, and new forms of political intervention were designed to distort, disrupt, and alter the way politics was routinely conducted. The high standards of rational decision making quickly dissolved under the eyes of the empirical observer, yielding their place to a mixture of power games, arduous negotiation processes of political bargaining, and recourse to already institutionalized conflict-solving mechanisms, such as the courts. While nothing in this is surprising to political scientists, it came as a surprise nevertheless to those who had thought that scientific advice was exempt from these ordinary forms of political rationality. When confronted with scientific advice and expertise, the policy process did not display the rationality expected. Scientific expertise was treated like any display the rationality expected. Scientific expertise was treated like any other input into the political process: as a political resource to be used by both sides, negotiable, and not necessarily "true"; in any case not endowed with higher political credibility than other inputs.

The second outcome is related to the first. IT underlines the inherent difficulty in reconciling the idea of scientific knowledge, generated in accordance with its methodological canon of objectivity and intersubjective validation, with demands of popular participation. What can be shared to a certain extent -"popularized" as the term has it- comes after scientific facts have been established and a body of knowledge validated. It is the diffusion of knowledge and, to some extent perhaps, its application that can be opened to public participation, but not the process of producing and validating scientific knowledge as such. Yet, in the public contestation phase, the objective findings of facts, their precondition as well as political consequences, were challenged. Thriving on the open disagreement of experts in public, a more transparent model of science for public policy was proposed, an adversary system that would allow for some kind of representational system of comparing scientific findings and methods of arriving at them. By juxtaposing experts and counter-experts, each chosen as trustworthy from the opposing parties, science was to become more democratic. Underlying such a proposal was of course the expression of a deeply-seated distrust of science functioning as an objective enterprise and standing above vested interests. In the public contestation, science was charged with taking sides with other powerful interest groups in society and therefore discredited as not being truly for public policy.

The other two changes affecting the dominant conception of science for public policy arose out of internal reflection and critical evaluation, notably through sociological studies of science. They show science not to be as neutral, objective, and free of social interests as the positivistic ideal of science affirmed for a long time, and claim that all scientific knowledge is socially constructed and negotiated[14]. Scientists were shown (in their own accounts of how they arrived at results) to oscillate between a usually informal context of contingency, in which they admit the uncertainty and provisional nature of the knowledge in question, and an empirical, formal context in which they justify the conclusions reached by emphasizing solely the certainty and absoluteness of the results they obtained[15]. Both of these themes represent revisions of the official model of science, the standard model confirmed by the public rhetoric of science. Although the critical dismantling of some of its features came from inside science, so did public controversies throw open the not-so-objective sides of objectivity and add the weight of context-dependency to the process of scientific inquiry. Among others, Brian Wynne has noted that it is important to see clearly that such criticisms and invitations to self-reflection are not to be taken as an all-out assault on science; nor is it a question of deliberate bias and wilful distortion on the part of scientists that needs to be publicly exposed. Rather, the all-pervasive message of such studies and detailed critiques is to make a much more general point: that the definition of a scientific problem is never isolated from the political context in which it occurs, nor can political implications be completely eliminated from the course of the analysis and policy conclusions derived only an the end[16]. Put in another way, I would add: we have to recognize and accept that all scientific analyses tied to a given policy context anticipates and reacts to the often unstated assumptions of policy outcomes. The use of concepts, the substantive implications of methodological procedures, the utilization of any kind of data cannot but be impregnated with different policy meanings. To claim anything else would be utterly naive and could not be upheld in the face of overwhelming empirical evidence to the contrary. How to utilize this knowledge for better public policy purposes is, however, still another matter.

In the period of public contestation and its aftermath, science for policy has been turned into the question of "science for whom?" While the policy arena has been potentially enlarged by a wider public that wanted to be heard, the lessons to be drawn from the demystification of the over-rationalized political process and the over-rationalized image of the internal workings of science are by no means clear. If we admit that policy-prone types of scientific analysis inevitably bear the marks of their contexts of justification, of contingency, and of political relevance; if we admit that the informal process of scientific reasoning, of the utilization of data, and their interpretation include much stronger doses of intuitive judgment, implicit values, and tacit procedures of persuasion -are we set on a course which leads not directly to hell, but to something akin, namely scientific relativism? Or, as many scientists (who still uphold the ideal of no science-in-public) would maintain, would a greater degree of honesty and modesty about the internal workings of the scientific process lead only to a further decline in public trust or increase public apprehensions, perhaps wilfully distorted even further by the media? Is there a way out from haughty retreat behind a formal position and from apologetic relativism alike?

4.4 Between orthodoxy and reformism

The orthodox response has been to reassert the traditional separation of science from public policy, arguing that only then can it be *science* for public policy. Similar statements abound in the policy field dealing with risk analysis, risk assessment, and risk management. A recent study prepared by the National Research Council of the US Academy of Sciences makes an explicit distinction between risk assessment and risk management: risk assessment is to be based on scientific judgement alone and has to find out what the problems are; it should therefore be protected from political influence. Risk management, on the other hand, is defined as the process of deciding what to do about the problems. It involves a much broader array of disciplines and is aimed towards a decision about control[17]. Perhaps more clearly than other policy studies, risk analysis has been confronted with the problematic situation that is inherently at the heart of most of them: while the intention is to provide as clear and careful a basis for action as possible by diligent scientific scrutiny of the hazards that can be subject to analysis, the selection and implementation of intervention measures generally involve balancing scarce resources, political goals, changing social values, and sometimes a somewhat unpredictable public opinion[18]. Another study published by a group of the UK Royal Society, equally devoted to methods and approaches to risk analysis, reached a different conclusion in which the whole process, including risk estimation, risk evaluation, judgments on acceptability, and taking account of public opinion, is referred to as risk management[19]. The respective roles of these two parts of the process are treated differently.

The chances for a successful application of the relativistic strategy are even slimmer. Not only is relativism a highly contested philosophical position within the theory of science[20], it has few, if any, friends among practising scientists. Even if we would leave aside the deeper philosophical issues and concentrate on a reformist plea for greater public openness about the internal side of science in which subjective judgements have their place, uncertainty abounds, and room is even made for errors -would this alone provide a better basis in the face of pressure for political action when confronted with incomplete and uncertain scientific knowledge? Although the public image of science is in urgent need of correction in the reformist vein, no miracles can be expected from this strategy if nothing else changes.

This takes us back to the questions raised at the outset of this chapter. If science only creates the conditions in which a society develops solutions, we may ask from a sociological point of view which kind of solutions are likely to emerge. If science itself takes proper notice of the increasingly recognized realm of uncertainty, due not only to the human condition of ignorance but to the knowledge gained about the interacting secondary and tertiary effects of scientific and technological interventions in the natural and social environments, the conditions are created for science *and* society to develop new kinds of solutions. On the epistemological side, this can be an intellectually exciting venture; for the policy process it might reveal some unexpected results.

So far, the historical conditions have favoured one particular type of solution: the utilitarian-instrumental one. Utilitarian solutions have pressed for the increased applicability of scientific knowledge, for its industrialization and more efficient organizational forms, and for its relevance to continued innovation. The concomitant societal mechanism aiming for the distribution of the surplus thus created, for motivation of the work force, and for the smooth functioning of societal institutions has been an instrumental type of rationality concerned only with efficient and hierarchical means-ends procedures that have become the guiding principle of how social affairs are conducted in the industrialized West. Yet, we have also come to realize recently that the conditions created by science and technology have increasingly cast doubt on the adequacy of these solutions as a guide for policy. The discussions about accelerated economic growth in the face of environmental damage and the threat to the overall balance between nature and man have been only one facet of growing uneasiness. Discussions within the scientific community on how to cope with uncertainty under the outside pressure for action have underlined the limits of the utilitarian-instrumental solution.

4.5 The rise of the managerial conception of science for public policy

The utilitarian-instrumental solution allowed for a clear-cut separation of science from policy while maintaining at the same time a strong (utilitarian) link of science for public policy, based on a means-ends relationship. While the production of scientific knowledge needed its autonomous space, it was assumed that it would lead more or less automatically to its social utilization since this was the in-built direction for scientific technological development to take. Steering clear of too-close a contact with the political system, "not meddling in politics", science became closely enmeshed with the industrialization process and its aftermath.

Science in now confronted with new demands from the political process. As with the industrial system, the question is not so much one of direct influence or control. The scientific system has guarded surprisingly well the core of its institutional autonomy. It was at the height of industrialization in the latter part of the nineteenth century that major industries in Europe became science-based, and the split between basic and applied science was successively introduced. I see something similar occurring today, with science yielding to the powerful and all-pervasive political context that demands new scientific solutions for dealing with problems that science and technology have helped to create. An institutional split -which is also epistemological, concerning methodologies, substantive content, and professional self-understanding alike- is likely to occur within the sciences- between a public policy branch and an academic branch. But there is no ready-made kit of tools and recipes, of techniques, nor computer simulation models which can easily be drawn upon to fill the knowledge gap. Rather, the epistemological and practical basis for this latest branch in the differentiation of the sciences is yet to be created. In order to be successful, it has to have a strong epistemological tradition within at least some of the sciences themselves; it has to hold out the promise of conceptual power and clarity and, at least, a methodological armoury that is adequate for the types of problem to be addressed. In short, it has to embody a vision of being able to meet the demands of the policy process without giving up its strong claims to institutional autonomy from direct political interference. In order to keep its position as monopoly holder of the most cherished type of knowledge and to be trusted by the public, confidence in its impartiality has to be restored. These criteria are met by a new conception of science for public policy which I call the managerial conception of science.

The development of the managerial conception occurred gradually and on several levels. At the height of environmental concerns, when the limits of growth and exploitation of natural resources became a newly perceived part of reality, resources were suddenly seen to be finite -to be managed for the interest of all. When technologies were threatening to get out of hand and in urgent need of new kinds of control, we started to speak of managing them. When it became clear that the new problems created through scientific-technological interventions, with their unknown, unintended, yet potentially harmful effects, could not be solved in the accustomed way -if ever at all- we switched in our rhetoric from solving problems to managing them. This is a reasonable adaptation to a new situation in which too many variables were interacting under highly uncertain temporal conditions and in which the resilience or robustness of systems had yet to be determined empirically and theoretically. The thought of management comes easily to systems thinking, as this is one of its more precisely defined roots.

The managerial conception of science for policy also contains an implicit plea for shared responsibility at a time when individual responsibility has lost all ground in the modern organization of science. It is no coincidence that it alludes to a corporate style: management of problems which cannot be solved; management of uncertainty rather than a quick and unfounded (irresponsible) hope that it disappear quickly. This contains an appeal to a multi-levelled hierarchy of responsibility adequate for the new kind of situation we era facing. In contrast to a notion like "muddling through" which Charles Lindblom proposed, with very moderate success, to explain the political process, the scientific management of problems proclaims a relatively high degree of control in the face of a sea of external uncertainties. It contains the promise of exploiting new opportunities, should they arise, and of ways to "identify and carry out actions that will allow us to change the rules of the game"[21]. In short, management, and especially scientific management, is a respectable, orderly procedure with a high degree of success in economic life, particularly within large-scale organizations. It implies a certain type of rational behaviour since it is a goal-oriented, but also takes account of unavoidable constraints. It has a formal and an informal side, as every student of organizational behaviour knows and good management is apt to utilize both to the fullest. Contrary to the political model of accountability, defined as the electorate in Western democracies, managerial accountability rests on the assumption of a built-in hierarchical structure of duties and liabilities which is only ultimately responsible to a distant and abstract entity (the "owners") who are not supposed to interfere. Thus, one of the strong appeals of the managerial model over a kind of political model lies in the high degree of autonomy it promises to the managers -in this case, to scientists. While it has remained problematic to defend the autonomy of science in the face of its role in the political process, the managerial conception promises a way out: while not denying the need for a built-in system of responsibility, its exact nature remains shrouded in a veil of competence in the double sense of the word; competence of those who are capable to handle scientific policy matters and of those who are officially charged with handling them.

The new conception of science for public policy -as distinct form academic science research- reduces the old question of how to maintain the boundary between science and public policy to irrelevancy, since by definition scientific management of policy problems stands above the need to protect science from political intrusion. It has all the evocative power of a new mediating institution and of a new social invention in the face of otherwise unsurpassable contradictions. It is an elegant solution and I predict that it will work successfully. It can incorporate the orthodox response and the reformist strategy described above: the former by interpreting the protective line being drawn between scientific fact-finding and political decision-making as being merely an administrative procedure; the latter by proclaiming greater honesty about inherent biases in the way science works as being part of the informal side of the management process.

The new ethos of science for public policy will be that of scientific managers, and good management is for the sake of the company. The only drawback I see is the question that remains open: who is the company and who controls it?

Notes

1. Reprinted with permission from H. Brooks, Ch. Cooper (eds). *Science for public policy*. Copyright 1987. Pergamon Press Ltd.

2. Brooks, H.(1981) Some notes on the fear and distrust of science, in A.S.Markovits and K.W.Deutsch, (Eds), *Fear of Science-Trust in Science* (Cambridge, USA: Oelschlager, Gunn and Hain Publishers).

3. Ravetz, J. R. (1985) Uncertainty, ignorance and policy, in Brooks, Cooper (eds.) 1987 *Science for public policy*, Oxford, Pergamon Press.

4. Brooks, H. (1981). *op.cit.*

5. Mendelsohn, E. and Nowotny, H. (Eds.) (1984) *Science between Utopia and Dystopia: Yearbook in the Sociology of the Sciences*, Vol.8 (Dordrecht, The Netherlands: Reidel).

6. Majone, G. 1981 Short... of the policy science approach to the analysis of the public sector, in F.X. Kautmann, G. Majone, and V. Ostrom (Eds) *Guidance, Control, and Evaluation in the Public Sector* (Berlin, FRG: Walter de Gruyter).

7. Wildavsky, A. (1979) *Speaking Truth to Power* (Boston, USA: Little, Brown, and Co.).

8. House, P. (1982) *The art of Public Policy Analysis* (Beverly Hills, USA: Sage).

9. Wittrock, B. (1983) *Policy Analysis and Policy-Making: Towards a Dispositional Model of the University/Government Interface*, Report No. 29 (Stockholm, Sweden: University of Stockholm, Sweden, Group for the Study of Higher Education and Research Policy).

10. Coates, J. (1978) What is a public policy issue? in *Judgements and Decision in Public Policy Formulation* (Washington, DC, USA: American Association for the Advancement of Science Selected Symposium 1) pp.34-69.

11. Schneider, J., Stevens, N., and Tornatzky, L. (1982) Policy research and analysis: an empirical profile, 1975-1980, *Policy Sciences*, 15, 99-114; Brunner, R. (1982) The policy sciences as science, *Policy Sciences*, 15, 115-135. See also Brewer, G. and de Leon, P. (1983) *The Foundations of Policy Analysis* (Homewood, USA: Dorsey).

12. Doughty Fries, S. (1983) Expertise against politics: technology as ideology on Capitol Hill, 1966-1972, *Science, Technology, and Human Values* (Spring).

13. Nowotny, H. (1984) Does it only need good men to do good science?, in *Science as Commodity* M Gibbons and B.Wittrock (Eds) (London, UK: Longman).

14. A good sampling of the literature can be obtained in *Social Studies of Science*.

15. Mulkay, M. (1983) Scientists theory talk, *The Canadian Journal of Sociology* 8 (2, Spring).

16. Wyne, B. (1983) *Models, Muddles and Megapolicies: the HASA Energy Study as an Example of Science for Public Policy*, Working Paper WP-83-127 Laxenburg, Austria: International Institute for Applied Systems Analysis).

17. National Research Council (1983) *Risk Assessment in the Federal Government: Managing the Process* (Washington DC, USA: National Academy Press); Ruckelshaus, W. P. (1983) Science, risk and public policy, *Science* 221, 1026-1028.

18. Coppock, R. (1983) *The Integration of Physio-technical and Socio-physical Elements in the Management of Technological Hazards*, mimeo (Berlin, FRG: Science Center, International Institute for Environment and Society).

19. The Royal Society (1983) *Risk Assessment. A Study Group Report* (London, UK: The Royal Society).

20. For a glimpse of an ongoing debate see Roll-Hansen, N. (1983) The death of spontaneous generation and the birth of the gene: Two case studies of relativism, *Social Studies of Science*, 13, 481-519.

21. Clark, W. C. (1980) *Witches, Floods, and Wonder Drugs - Historical Perspectives on Risk Management*, R-22 (University of British Columbia, Canada: Institute of Resource Ecology).

Managerial Science, Inc. Revisited

Looking back in retrospect to an analysis written almost six years ago, my approach is a mixture of inevitable curiosity and distance. Curiosity, since it is always of interest to observe how the diagnosis of yesterday fits into the frame of analysis of today and distance since it is obvious that many profound changes have occurred since. Tucked somewhere between this curiosity and the immunizing distance there is also room for surprise.

My first reaction was that the analysis reprinted above holds up quite well to the test of time. Obviously, the trends I and others could observe then were clearly under way and could be laid bare to shared observation and comment. The old conception of science for public policy was ostensibly giving way to something new which, with a tinge of irony, I called the managerial conception of science: from now on, I wrote, whatever problems could not be solved, would have to become managed. And I went on with an almost lyrical tone to describe what this new conception of science would entail: like any monopoly holding firm under competitive pressure it would set up its new marketing division. Science Inc. would reach out and preempt any competitive assault. While promising increasing participation and attendance to users'needs, it would subtly scale down those expectations it clearly could not fulfil. While maintaining its monopoly of technical competence, Science, Inc. would give itself the guise of new and more modest image in public, emphasizing that uncertainty existed but reassuring the public at the same time that everything was under control. But Science. Inc. through its many subsidiaries in form of experts would also be ready to open up new negotiating space whenever public controversy demanded so. Like with any good management, the top would emphasize long-term planning and vision over short-term and attempt to foresee to the maximum extent possible where future trouble might arise. Technology assessment, different forms of evaluation, setting up high level advisory committees and other mechanisms should see to that. In short, the answer I saw forthcoming to the embattled old conception of scientific rationality feeding into the public policy process would be an essentially technocratic version in a new guise: by widening the negotiation space and by setting up new intermediary institutions and mechanisms destined to mediate between the policy process and science, a good management conception borrowed from the business world would seem convincing enough to stave off what I regarded the real question: who would be in control in defining which problems were scientific, where the lines would be drawn between purported facts and purported values. Whose science would be selected by whom for being channelled into public policy when public controversies involving technological risks would not be assuaged quickly but were here to stay. And would those running Science. Inc. begin to understand that the public unrest was not simply anti-scientific, but legitimate and profound concerns about the future directions society would take and how their own lives would be affected by them? Behind the ironic tone I can detect now a profound note of concern that the real issues would not even be addressed.

So what has happened since? The demise of the old model of rationality inherent in the policy process and in the mechanisms through which the purported rationality of science would be fed into it, has proceeded with surprising speed. Indeed, upon rereading my text some of the characteristics of the old model do not seem just a few years away, but decades. The language of describing rational planning processes just as much as the faith that once upheld them, seem completely out of place and belonging to a by-gone age. While their ongoing demise was obvious then, the acceleration with which it has taken place fills me with surprise. It is related to the profound changes that have occurred since between the room given to market forces and the state. Especially in Europe, the all powerful welfare state in the glorious three decades in which it reigned supreme after the war, brought with it not only powerful and efficient alliances between modernizing elites and social scientists and others eager to advise policy-makers, but also was built upon the belief into rational planning and an as yet largely unclouded vision of further human betterment and progress brought about by science and technology. Not only have these political alliances fragmented, and the former pervasive consensus into a "societal project" of major proportions been shattered. The once powerful State has lost control and with it once possible regulatory mechanism have been rendered unfeasible. Political majores ties are difficult to set up almost everywhere, demagogical elements are pervasive. A process of desolidarisation has set in. Technological innovation has become the prime mover of economic competitiveness between nations. Public expenses for science and technology are legitimated by bluntly pointing to the links between technological innovation, international competitiveness and private wealth. After the fall of the Communist regimes, science and technology have emerged in Eastern Europe as deeply implicated with the old totalitarian order. As a reaction new waves of anti-science are on the rise. In the United States where the state never played the central role it did in continental Europe, other dispersive forces have come to the fore, strongly aided by an adversary legal political culture that relies heavily on court decisions. Creationism won some local victories and "junk science" was able to enter the courtrooms. While the percentage of legislation containing a scientific and technical dimension has increased manifold, the mechanisms of regulating disagreements and conflicts that ensue, have apparently lagged behind.

Undoubtedly other factors have contributed to the demise of the old conception of science for public policy. The public demand for participation in decision-making regarding future technological and scientific developments that were seen as carrying risks, was here to stay, regardless of how participation would be defined in the end. It became reinforced through processes that were built upon the increasing funding needs of the research system which led to a greater demand for public accountability of science and technology. More recently, scientific fraud (itself closely linked to intensified pressure for publications, especially in the life sciences) and alleged mismanagement of public research funds, have become highly visible and hotly debated issues in the United States. In many ways one might plausibly argue (as I have done elsewhere, Nowotny, 1990) that science has become much more like other institutions in modern society: in its organisation forms and work setting, in its moral standing and the norms held by its members, and in its affinity towards (some would say corruption by) economic and political influence. From this it is easy to make a more discomforting conclusion: if science is not so special, it has no grounds upon which it can claim privileged status as far as its funding base and political support is concerned. It is but a short step to being treated like any other institutions that functions to serve society. And it is but yet another short step to be affected by this process of "societal normalization" in its claims towards a higher form of rationality and consequently towards higher cognitive status and authority.

In de-mystifying science and technology and in deconstructing what has become widely accepted as a "social construction" with inbuilt social negotiation processes, social studies of science and technology have played an epiphenomenal role. While these studies are vitally important in elucidating the social nature of science and technology and in showing how they are deeply embedded in society, it is patently too simple and moreover displays ignorance of how the social sciences work when they are seen merely as attempting to subvert the authority of the natural sciences while offering little in return. While it is understandable that natural scientists react with irritation to what they see as the subversive and demystifying result of much social analysis, it would be grossly overestimating the potential impact of any social science analysis to attribute the actual process of legitimation of science to such analysis alone. If we look for the power of ideas and their cultural influence in changing the general outlook of a society, then we have to take into account much wider currents flowing in the cultural sea. The decisive turn is one from a modern society to one that believes itself to have moved beyond this programme and its so called postmodernity. Postmodernism is far from being merely a cultural fad or an intellectual pass-time for philosophers. Recently, Thomas Hughes, the eminent historian of technology, spoke about "post-modern technology". What he meant were among other the radically altered conditions of production and intellectual and social creativity under which technologies are being created today. This relates to modes of financing as much as to styles of corporate management, to the interlinkages of science and technology with contexts of their potential use which need to be taken into account and built into their making right from the beginning (Hughes, 1992). Post-modernism is built upon deconstruction of any central authority, it entails fragmentation and the loss of any central perspective. But it also enables new forms of local action and allows for contingencies, it entails entirely new possibilities of 'just connect" through multiple configurations through which creative new interlinkeages are being networked. Far from being only a bundle of ill-defined concepts and methods, it has seeped into the social fabric of society, into the relationships of a former centralized locus of power and authority and those who are connected to it. In essence, postmodernism is an answer that has evolved in response to the greatly increased societal complexity. Is it surprising that science as a central authority in our highly industrialized societies is also affected by it?

So far, the response on part of the natural scientific establishment has been one of covert bewilderment. The managerial model of Science, Inc. has been put into place, yet it does not prove quite as effectively as it was supposed to be. Sheila Jasanoff in a recent contribution detects a definite technocratic bent in the ongoing renegotiations of the boundary separating science and politics. She sees a more general trend emerging in the US towards reducing the power of lay perspectives to influence the direction of science and technology policy, partly as a reaction towards the role of experts in the courtroom, where unrestricted access of experts to the jury is seen by segments of the policy community as an invitation for "bad science" to crowd out the "good". She points out that both scientists and the state have a stake in representing results as "science" so as to protect them against renewed destruction (Jasanoff, 1992). Another scholar, Yaron Ezrahi, who has attentively observed science in the framework of transformation of contemporary democracy, is convinced that the growing public distrust of science and technology is more a symptom of the changing conception of politics than of science and technology per se. Yet this changing conception has radical consequences also for science and technology. While major economic, military, or social crises are likely to revive the rhetoric of realism, the criticism of "pleasing illusions" and the appeals to science and technology, Ezrahi believes that for the time being, "stagecraft" as the art of eloquent, edifying and politically effective gestures is the supreme technique of statecraft. But he also cautions that the "descent of Icarus", the delegitimation of grand social and political engineering and the decline of instrumental rationality in the context of public affairs, does not necessarily represent a return to darkness. Postmodern politics and postmodern science will have to face their own respective limits (Ezrahi, 1990).

Perhaps facing their limits while renegotiating their boundary is nowhere as manifest and urgent than in the field of the environment. The very term of "science for public policy" ceases to be meaningless there when science is taken to imply a relatively closed, self-contained set of knowledge which is supposed to provide guidelines for action. The complexity of the problems, their truly global nature and their "wholeness" raises much more fundamental questions about scientific determinacy, uncertainty and the relationship of scientific knowledge to policy advice and action. In a sense, environmental problems defy any one-way approach to solutions. They call for "management", yet the style and form of management itself has changed and assumed some definitely post-modern features. Multiple local actions, all embedded into economic and social arrangements and global environmental effects are interlinked in almost intractable ways, while global action, the aim of reaching international consensus and binding agreements, is filled with obstacles that deliberately thrive on scientific uncertainties. And while it can be said that science-based industry and industrialized science got started when academic science was moved into specially set-up industrial research labs in the latter part of the 19th century, today we can see the necessity of moving the environmental sciences out into society. This makes them open to new claims of societal access. A whole new set of issues is awaiting recognition and some form of incorporation into science: environmental rights of citizens, a new "contract" with nature, intergenerational rights, communal rights for common property, such as the atmosphere, and the oceans, the setting up of environmental codes of conduct and definition of the responsibilities of scientists. Science is challenged to become "vernacular": conversant with mature citizens and willing to accede to certain of their legitimate demands. It is asked to share knowledge and information, the process of monitoring and its results. The management of sustainable development calls for a new kind of cooperation between natural science and social science, but it is a management which can only hope to succeed if it does not fall into either of the two old traps: to become science infiltrated by politics or to become science aloof from society.

Thus, the managerial model of Science, Inc. is up for some hard testing. It will be tested for the sincerity with which it is open to citizens' demands, even if openness will not necessarily bring with it political consensus. It will be tested for the robustness and adaptiveness of its intermediary organizations which spring up all over in attempts to renegotiate boundaries and to make transactions between science and politics more productive. Will the new species of hybrid experts working in these intermediary spaces be able to develop an ethos of their own, one that will enable them to speak to each other as honest brokers? If one of the answers of the postmodern managerial style is pluralism, if it is true that only "divided we stand" (Schwarz and Thompson, 1990), that we have to accept our differences in beliefs and outlooks and utilize them consciously as resource, then it will be necessary to develop a political and scientific culture based upon a widely shared understanding and acceptance of that pluralism. In the end, there cannot be only one solution, for this very pluralism implies that science and society will have to find - and accept - many different configurations in which their boundaries are redrawn. If there is a lesson to be learned, its sphere of relative autonomy and its claims towards a higher form of cognitive rationality only, if it is able to incorporate and accommodate sufficient elements of ongoing societal discourse and of the presently occurring societal structural changes. For science is no longer immune from society.

References

Ezrahi, Yaron (1990).
The Descent of Icarus. Science and the Transformation of Contemporary Democracy. Cambridge, Mass.: Harvard University Press.

Hughes, Thomas (1992).
"Modern and Postmodern Technology". Paper presented at the International Conference *'The social impact of science and technology: modernization of society between market and state'*. Genova, Badia di S. Andrea, 9 - 10 April 1992.

Jasanoff Sheila (1992).
"Pluralism and Convergence in International Science Policy". Paper prepared for IIASA 92. *An international Conference on the Challenges to Systems Analysis in the Nineties and Beyond.*

Nowotny, Helga (1990).
"Individual Autonomy and Autonomy of Science: the place of the individual in the research system. In: S.E. Cozzens et al (eds.) The Research System in Transition. Dordrecht: Kluwer Academic Publishers, 331 - 343.

Schwarz, Michel & Thompson, Michael (1990).
Divided We stand. Redefining Politics, Technology and Social Choice. Philadelphia, University of Pennsylvania Press.

5 THE EMERGENCE OF POST-NORMAL SCIENCE[1]

Silvio O. Funtowicz* and Jerome R. Ravetz**

* Institute for Systems Engineering and Informatics, Joint Research Centre, Commission of the European Communities, Ispra (Va), Italy.

** The Research Methods Consultancy Ltd., London, England.

5.1 Introduction

Few will still doubt that our modern technological culture has reached a turning point, and that it must change drastically if we are to manage our environmental problems. It may not yet be as widely appreciated that science, hitherto the mainspring of that technological progress, must also change. From now on its central task must be concerned with the pathologies of our industrial system; and this imposes new problems and requires new methods. These are the subject our of study.

The fundamental achievements of science, like those of all creative activities, have a timeless quality. The social activity of science, like any other, evolves in response to its changing circumstances, in its objects, methods and social functions. In the high Middle Ages, the independence of secular learning was established in the universities, removed from the monasteries; and the boundary between the sacred and private on the one hand, and the secular and public on the other, was set for European culture. The Scientific Revolution of the seventeenth century was one of the great intellectual mutations of mankind, and reinforced the growing hegemony of European civilization in the world. The nineteenth century saw the replacement of "natural philosophy" by science, the growth of subject specialties, the institution of a value-free Scientific Method, and the first career opportunities for scientists. Parallel to this came the consolidation of the science-based professions, with their own institutions and formalized social contracts. In the recent postwar period we have experienced the industrialization of science, with the growth in scale and capital-intensity of research and its intimate connection with technology and political power. Paradoxically, as science prospered materially, it was losing its ideological function as the unique bearer of the True and therefore the Good.

Now the global environmental issues present new tasks for science; instead of discovery and application of facts, the new fundamental achievements for science must be in meeting these challenges. Because of the very rapid changes in environment, society and science itself, and in their interactions, a general awareness of the new state of science has yet to be achieved. In this essay we make the first articulation of a new scientific method, which does not pretend to be either value-free or ethically neutral. The product of such a method, applied to this new enterprise, is what we call "post-normal science".

We adopt the term 'post-normal' to mark the passing of an age when the norm for effective scientific practice could be a process of puzzle-solving in ignorance of the wider methodological, societal, and ethical issues raised by the activity and its results. The leading scientific problems can no longer derive from abstracted scientific curiosity or industrial imperatives. They are thrown up by issues where, typically, facts are uncertain, values in dispute, stakes high and decisions urgent. When research is called for, the problem to be studied must first be defined, and this will depend on which aspects of the issue are most salient; hence political considerations constrain which results are produced and thereby which policy implications are supported. In general, the post-normal situation is one where the traditional opposition of 'hard' facts and 'soft' values is inverted; here we find decisions that are 'hard' in every sense, for which the scientific inputs are irremediably 'soft'.

5.2 Uncertainties in research related to policy

The concept of uncertainty is at the core of the new conception of science, for hitherto it has been kept at the margin of the understanding of science, for laypersons and scientists alike. Whereas science was previously understood as steadily advancing in the certainty of our knowledge and control of the natural world, now science is seen as coping with many uncertainties in urgent technological and environmental decisions on a global scale. A new role for scientists will involve the management of these crucial uncertainties; therein lies the task of quality assurance of the scientific information provided for policy.

The new global environmental issues have common features that distinguish them from traditional scientific problems. They are global in scale and long term in their impact. Data on their effects, and even data for baselines of 'undisturbed' systems, are radically inadequate. The phenomena being novel, complex and variable, are themselves not well understood. Science cannot always provide well-founded theories based on experiments for explanation and prediction; but can frequently achieve at best only mathematical models and computer simulations, which are essentially untestable. On the basis of such uncertain inputs, decisions must be made, under conditions of some urgency. Therefore science cannot proceed on the basis of factual predictions, but only on policy forecasts.

Computer models are the most widely used method for producing statements about the future based on data of the past and present. For many, there is still a magical quality about computers, since they are believed to perform reasoning operations faultlessly and rapidly. But what comes out at the end of a program is not necessarily a scientific prediction; and it may not even be a particularly good policy forecast. The numerical data used for inputs may not derive from experimental or field studies; the best numbers available, as in many studies of industrial risks, may simply be guesses collected from experts. (And who has expertise in choosing experts?). Instead of theories which give some deeper representation of the natural processes in question; there may simply be standard software packages applied with the best-fitting numerical parameters. And instead of experimental, field or historical evidence, as is normally assumed for scientific theories, there may be only the comparison of calculated outputs with those produced by other equally untestable computer models. Thus in this post-modern science of simulation, computer power allows articulation and flexibility to substitute for versimilitude and testing against an external reality.

In spite of the enormous effort and resources that have gone into developing and applying such methods, there has been little concerted attempt to see whether they contribute significantly to our knowledge or to the quality of our decisions. In research related to policy for risks and the environment, apparently so crucial for our well-being, there has been very little effort of quality assurance of the sort that the traditional experimental sciences take for granted in their ordinary practice. Whereas computers could in principle be used to enhance human skill and creativity by doing all the routine work swiftly and effortlessly, they have tended to become substitutes for thought and scientific rigour. Indeed, some distinguished scientists have questioned whether computer models should be used at all in the study of the global environmental problems. Thus the American mathematician S. Mac Lane describes 'systems analysis' as, "the construction of massive imaginary future"scenarios" with elaborate equations for quantitative "models" which combine to provide predictions or projections (gloomy or otherwise), but which cannot be verified by checking against objective facts. Instead, [such] studies often proceed by combining in series a number of such unverified models, feeding the output of one such model as input into another equally unverified model... Such studies as these are speculations without empirical check and so cannot count as science... (1).

In his defence of the field, N. Keyfitz reminded us that "many of the most difficult problems we have to face cannot even be precisely formulated in the present state of knowledge, let alone solved by existing techniques of science.... Such models, although unsatisfying to many scientists, are still the best guide to policy that we have.... (2).

In his reply, Mac Lane continued to doubt that the global problems should be tackled by making models "that in the first instance are not verifiable", and added, "problems are not solved and science is not helped by unfounded speculation about unverifiable models". His concluding comment was on quality assurance: to the effect that the research institution he was criticizing "does not appear to have an adequate critical mechanism, by discipline or by report review." (3).

To believe that the calculated outputs of untestable computer simulations should determine policies, is to indulge in the purest rationalistic fantasies, reminiscent of Leibniz or better of Ramon Lull. Indeed, we may speak of a new sort of pseudo-science, depending not on magic but on computers, which can be called GIGO ("Garbage In, Garbage Out"). This can be defined as a computational field where the uncertainties in the inputs must be systematically suppressed, lest the outputs become completely indeterminate. How much of our present social and environmental science belongs to this category, is an interesting and urgent question. Parallel to these computer-based pseudo-sciences are the computer-based pseudo-technologies. These based their appeal on a confusion between adequate computer graphics of an excellent technological system and excellent computer graphics of an imaginary technological system .

It is clear that the dilemmas of computer modelling in research related to policy cannot be resolved at the technical level alone. No one claims that the computer models are adequate tools; and yet nothing better can be provided by traditional science. The critics basically judge them by the standards of mathematical-experimental science, and of course in those terms they are nearly vacuous. Their defenders advocate them on the grounds that they are the best possible, without appreciating how very different are these new sciences of clean-up and survival in respect of their complex uncertainties, new criteria of quality and socio-political involvements. The need is for exceptionally dedicated efforts for the management of uncertainty, the assurance of quality, and the fostering of the skills necessary for these. Such skills will not be easily developed within the old framework of assumptions about the methods, social functions and qualified participants in the scientific enterprise.

The uncertainties in research related to policy are not restricted to computer models. Even the empirical data that serve as direct inputs to the policy process may be of doubtful quality. Their uncertainties are frequently incapable of management by traditional statistical techniques. As J.C. Bailar puts it:

> "All the statistical algebra and all the statistical computations are of value only to the extent that they add to the process of inference. Often they do not aid in making sound inferences; indeed they may work the other way, and in my experience that is because the kinds of random variability we see in the big problems of the day tend to be small relative to other uncertainties. This is true, for example, for data on poverty or unemployment; international trade; agricultural production; and basic measures of human health and survival. Closer to home, random variability - the stuff of p-values and confidence limits, is simply swamped by other kinds of uncertainties in assessing the health risks of chemicals exposures, or tracking the movement of an environmental contaminant, or predicting the effects of human activities on global temperature or the ozone layer." (4)

Thus in every respect the scientific status of research on these policy-related problems is dubious at best. The tasks of uncertainty management and quality assurance, managed in traditional science by individual skill and communal practice, are left in confusion in this new area. New methods must be developed for making our ignorance usable (5). The path to this lies in a radical departure from the total reliance on techniques, to the exclusion of methodological, societal or ethical considerations, that has hitherto characterized traditional science. This is the challenge that has led us to develop the idea of post-normal science, as the sort of science that is appropriate to this post-industrial civilization.

5.3 Uncertainty, quality and values in science for policy

Any policy decision on global environmental issues will need to be made in the context of uncertainty, dependent on inputs of variable or even unknown quality. There is a growing concern among experts, politicians and the public about the uncertainties affecting data for major environmental issues, such as global warming. There seems to be no systematic solution to this problem; instead, uncertainty is manipulated politically, for accelerating or deferring major initiatives, depending on the outlook of the advocate. By contrast, the problem of quality assurance of information has been almost universally ignored. One reason for this neglect may be in the confusion between uncertainty and quality, and the naive belief that there is a straightforward relationship between them, high quality being equivalent to low uncertainty.

Hitherto the handling of these problems has oscillated between two extremes. At one end there are perceptive philosophical analyses about the relation of knowledge and ignorance (5,6), and of the general phenomenon of quality criteria as employed in the policy process (7). These provide a reflective understanding, but they cannot easily be translated into practical tools for quality evaluation of uncertain information. At the other extreme are the technical uncertainty analyses (8,9,10). and simple quality taxonomies (11,12). These combine classifications of sources of uncertainty, specific to each field, with mathematical formalisms that treat uncertainty as if it were an additional physical variable. It is small wonder that those who must cope with uncertainty in their work will generally ignore the whole subject in practice.

Whereas uncertainty is an attribute of knowledge, quality is a pragmatic relation between a product, or process, and its intended users. It can be defined as 'the totality of characteristics of a product that bear on its ability to satisfy an established use' (13). Uncertainty and quality are two distinct attributes, for information of lesser certainty may yet be of good quality for its intended function. An extreme case of this is provided by deforestation in the Himalayas; although the estimates of the per capita fuelwood consumption vary through a factor of almost a hundred, all serious studies agree that their numerical predictions imply that the problem exists and that its solution is urgent (14). An example of high certainty and very poor quality is provided by a prediction of a rise in the average temperature of the earth of 0 to 10°C over the next forty years due to the greenhouse effect. On a common sense basis, we may say that the true value is almost certain to lie within that range; but the climatic consequences in this range vary from the trivial to the nearly catastrophic. The prediction is nearly true by definition; its quality decreases accordingly because the statement approaches being analytical rather than synthetic, in other words, it tells us very little about the real world .

Thus there are inherent limitations to the reduction of uncertainty in this kind of research. There is no point in ecological modelling (for example) trying to emulate experimental physics in its control of uncertainty. Each field of practice has a characteristic grade of information (rather like hotels or restaurants in grading schemes) appropriate to its needs; within that grade, information may vary in quality (as with hotels), depending on how well its uncertainties are managed and hence how well the information fits its function as an input to a decision process.

In ordinary scientific practice, considerations of values are largely implicit; even if they are operative in the choice of problems, once the research is underway they are put in the background. However, they are always present as part of the framework of the research; the myth of "value-free" science can be sustained only by ignoring the routinely used statistical methods. In any genuine statistical exercise, the design must take account of the error-costs of the possible alternatives; thus no single test can optimize both selectivity and sensitivity (avoiding the errors of false-positives and false negatives). The choice, as expressed in numerical confidence-levels, reflects the background of values, realized as costs and benefits, which condition every experimental program.

When ordinary scientific practice does not provide conclusive solutions for its problems, the values become explicit in the assignment of rules of inference. The growing use of scientific expertise in the courts frequently reveals a mismatch between the traditional value-implicit rules of scientific inference and those appropriate in tribunals. Thus in the law courts, various special principles for controlling error-costs are invoked, including "balance of probabilities" and "burden of proof". Thus in the latter case the error-costs of convicting an innocent person are deemed to be higher than those of acquitting the guilty, at least in the Anglo Saxon tradition. Tribunals of inquiry provide an illuminating case of bridging between the two approaches and their appropriate conceptions of value and error-cost. In the Black enquiry on the excess child leukemia cases in the neighbourhood of the Sellafield nuclear reprocessing plant, the Scottish concept of a 'not proven' verdict was explicitly applied for the possible cause of the excess leukemias (15).

In problems of risks and the environment, the value considerations in scientific practice may be quite explicit. For a classic example, we may consider the statistical design of a program for testing defective copies in a large shipment. If it is of apples, then a bad one spoils only its barrel; but if it is of landmines, a premature explosion can take the whole neighbourhood with it. The relative costs of false-positives and false-negatives are very different in the two cases. This example also serves to illustrate the factor of "dread" which is an important dimension of public perception of novel risks like nuclear power and release of genetically engineered organisms.

An integrated approach to the problems of uncertainty, quality and values has been provided by the NUSAP system. In its terms different sorts of uncertainty can be expressed, and used for an evaluation of quality of scientific information. NUSAP enables us to make the distinction between the sources and the sorts of uncertainty. Classification by sources is normally done by experts in a field when they try to comprehend the uncertainties affecting their particular practice. But for a general understanding, we have to distinguish among the technical, methodological and epistemological levels of uncertainty; these correspond to inexactness, unreliability and "border with ignorance", respectively (16).

Uncertainty is managed at the technical level when standard routines are adequate; these will usually be derived from statistics (which themselves are essentially symbolic manipulations) as supplemented by techniques and conventions developed for particular fields. The methodological level is involved when more complex aspects of the information, as values or reliability, are relevant. Then personal judgements depending on higher-level skills are required; and the practice in question is a professional consultancy, a 'learned art' like medicine or engineering. Finally, the epistemological level is involved when irremediable uncertainty is at the core of the problem, as when modellers recognize 'completeness uncertainties' which can vitiate the whole exercise, or when 'ignorance-of-ignorance' (or 'ignorance-squared') is relevant to any possible solution of the problem. In NUSAP these levels of uncertainty are conveyed by the categories of spread, assessment and pedigree, respectively.

There is no strict correspondence between these conceptual sorts of uncertainty and the sources we mentioned that are derived from practice. All data are affected by inexactness, and all computer models by ignorance. But all data exist within the framework of structures of concepts and procedures for their production, and of theories for their interpretation; hence the higher levels of uncertainty are relevant to their evaluation and use. Therefore, we may say that ignorance is a part of data uncertainties. Similarly, the lowest level of uncertainty, inexactness, occurs in computer models, through the use of numerical analysis techniques which unavoidably involve rounding-off and other approximation methods. Hence, the two approaches to the classification of uncertainty are quite distinct. A taxonomy based on sorts of uncertainty, like that of NUSAP, enables the construction of a general tool for the explicit communication of quality and values of the kind that appear in global environmental issues and policy related research.

5.4 The discovery of uncertainty in science

Now that global environmental issues provide the most challenging problems for science, uncertainty is moving in from the periphery, one might say the shadows, of scientific methodology, to become a central, integrating concept. This is the culmination of a process that has extended nearly a century, after almost three centuries of dominance of a triumphalist ideology of science in whose terms uncertainty was vanquished. At each stage of the process, the very successes of science have, in dialectical fashion, raised problems whose insolubility revealed radical uncertainties.

The first major crisis was in mathematics, the most technically sophisticated and ideologically sensitive field of science. By the turn of the century, contradictions at the logical foundations of mathematics were revealed, and within three decades their insolubility had been demonstrated by Gödel. In 1905 came Einstein's new physics, displacing inherited presuppositions, as of absolute space and continuity. Later developments produced the theoretical uncertainties of quantum theory, as in the Heisenberg theory and wave-particle duality. Notwithstanding such philosophers' worries, scientists and technologists pressed on to new heights of achievement, culminating in the atomic bomb with its revival of the ancient moral uncertainties of "knowledge too powerful to be revealed".

Uncertainties in knowledge reasserted themselves in practical fields with the development of technologies that are so novel and complex that the traditional skills of industrial safety assurance are inadequate. In such enterprises, (notably civil nuclear power), it was science rather than engineering that experienced a crisis, for it was scientists rather than engineers who took credit for the achievements and created the public expectations of a brave new technological world. When the new problems arose, scientists were not well prepared for them, because generations of a sheltered existence inside universities and academies had alienated them from the world of practice that was familiar to Leonardo da Vinci, Galileo and even Newton. In the recent past, those scientists who worked for clients, such as some chemists and statisticians, were socially marginal. The core sciences were those closest in all ways to the traditional humanities, where prospects for employment were the least, and entry favoured those with patrons or wealthy parents. In the context of this isolation from societal concerns, the ideology of a science that was value-free and ethically neutral, and the myth of a 'disinterested' scientist, could flourish, and indeed could foster the special sort of excellence of academic science. As a social activity, it operated largely autonomously and informally in the three crucial functions of selection and definition of problems, evaluation of results, and management of intellectual property. There was a largely implicit "social contract of science", whereby science indirectly provided all sorts of societal benefits, in return for the largesse from society.

In such circumstances, those with a scientific training whose research was directed by an employer (private or State) rather than by an informal peer-community, could be envisaged as doing the same sort of thing as their more fortunate academic colleagues, though with rather less freedom. In historical experience, such applied science provided (along with teaching) the bulk of the career opportunities for science graduates. It was a hybrid activity, very similar to core science on the cognitive side, while on the institutional side lacking many of the advantages and amenities of academic science. Problems were chosen by superiors with a view to applicability, solutions evaluated similarly, and intellectual property belonged to the employer. Historically this applied science was quite distinct from engineering, having different tasks and a different social organization.

During the Second World War all this was changed, as the world's most famous scientists congregated in the military research laboratories of the United States and Great Britain, solving problems of military engineering with their scientific approach. (In addition to the obvious case of the atomic bomb, there were the solution of the 'queuing' problem by Wiener and von Neuman, and code-breaking by Turing and his colleagues in mathematics and logic). In the early postwar period, these scientists and their students were involved in the conception and development of the most advanced technologies (such as electronics, nuclear power and space), and therefore these great engineering projects were generally conceived in the image of science. Science itself was being transformed, becoming 'big' or industrialized, with gigantic research projects, so that the distinction between science and engineering was further blurred. The basic distinction between scientific validity and engineering feasibility was overlooked, particularly in the euphoria of the 1950's. Hence the characteristic problems of the development and regulation of these new technologies, which could have been familiar to any engineer, were for a long time completely unnoticed. And the lack of scientific exactness in the calculations of risks, which every engineer knew and managed through 'engineering judgement' and 'good practice', came as an unwelcome surprise to the scientists who tried to apply the laboratory style to these new and difficult problems.

The politically sensitive problems of industrial risk assessment were the first to expose the conflict between the dominant self-image of science and the demands of a new form of practice. Only a few years separated the naive confidence of the Rasmussen report on reactor safety (17) from the patent bewilderment of the experts during the weekend of the Three Mile Island accident. During the repeated crises of the '70s and '80's over industrial and environmental risks, fields of technology which had previously legitimated themselves as sciences suddenly revealed their fallibility. The old conception of science as the guarantor of the Good and the True had passed into history.

Looking back on this period, we can see it as one where reality suddenly broke in and left the official reassurances and the scientific rhetoric emptied of significance. The ongoing crisis of a shortfall in recruitment to science by younger generations could be a case of what happens when a social system loses its claims to idealism. The science of the postwar period promised so much: the conquest of poverty through cheap nuclear power; the conquest of disease through cancer research; and the conquest of 'the last frontier' through space travel. In retrospect so much of that effort seems misdirected; what new cause can provide the challenge that will stir the imagination of the young? In one sense it is already there with 'the environment'; but if the environmental problems are managed in the same old way, who will want to find a vocation there? The old fragmented academic science, as the model for applied science in technological problems, has given way to professional consultancy. We shall analyze the strengths and limitations of this form of practice, as a preliminary to the explanation of the post-normal science that is appropriate for the present age.

5.5 Professional consultancy

The fields of professional consultancy had long been familiar with the problems of risks and responsibilities. They had developed appropriate forms of social organization; their "colleges" would admit and also expel members, and operate codes of professional etiquette and ethics. Their social contract did not involve the same degree of informality and autonomy as science. In return for a guarantee of good practice, the professions were allowed to police themselves, thereby providing their members with legal immunity in all but the most extreme cases. They could not pride themselves on being members of the "Republic of Letters", where intellectual property takes a very tenuous form; but they could take satisfaction from deploying their expertise for human welfare, and on occasion doing so in situations fraught with hazard.

Whereas the scientist's task is completed when he has solved a problem which in principle can function as a contribution to a body of knowledge, the professional's task involves the welfare of a client, and the science that is deployed for that is subsidiary to that goal. One way of appreciating the difference between the two vocations is in their degree of recognition of skills and judgements. In the traditional philosophical conception of science, these are irrelevant for the validation of results, however much they may be involved in the research process. Only a few philosophers of science (18,19) have accorded them any significance. Among professional consultants these are known to be paramount; indeed the fee structure reflects the knowing how to do a job (and being certified as such) more than its simple performance. Integral to a genuine profession is the prolonged apprenticeship, which may involve years of initiatory rituals of hard study or overwork. But the real rationale for these systems is the training in professional craftsmanship, which involves skills at the intellectual and behavioural levels. All this is in striking contrast to traditional science, where the single piece of supervised research, registered in a Ph.D. degree, has been deemed an adequate warrant for a lifetime career in independent research.

Further, the professional consultant can never escape from the tension between the scientific facts as established in the course of training, and the elements of a decision in everyday practice. For this practice involves the management of uncertainty of many sorts. The tasks as presented will (particularly in the challenging cases) correspond imperfectly to the idealized categories of formal education. On occasion, the professionals may have to cope with situations for which their training provides little or any direct guidance. Faced with such contingencies, someone inside a scientific tradition would tend to protect himself by qualifying his conclusions with many caveats and always demanding more time for further research. The professionals know that they cannot afford such luxuries. Responsibility for the consequences of decisions and actions, rather than the guarantee of the validity of conclusions, is their metier. (20).

Directly relevant to our present study is the difference between applied science and professional consultancy in the case of a decision which formally depends on the results of research, but where the stakes are high. In the traditional approach of science, even of applied science, such a consideration was deemed irrelevant. Scientists were to find the facts to the best of their ability, and that was the end of the matter. But in a real conflict of interests, any stakeholder can find ways of strengthening his position by criticizing the methodology of the other side's work. Because of the open texture of scientific argument (as distinct from logic or mathematics), such arguments can be prolonged indefinitely. In such a situation, the role of the professional consultant takes on another dimension. In this forensic context, the client's concern is less for the (contested) "facts" than for his threatened interests. The professional's responsibility for the wellbeing of his client can come into conflict with his own long-term interests and those of his profession. It is not only in such situations that the professional encounters ethical problems; and however imperfectly they have been incorporated into the training and institutional practice of professionals, there has never been any doubt of their existence. This is in striking contrast to the scientists, who until very recently were not even aware of having clients to whom they had ethical or societal responsibilities. In the present period, the degree of awareness of uncertainties and ethical problems which professionals have developed, is insufficient for the effective use of research results in the decision process.

We should not think that the professional consultants can confront some reality "out there" independent of a cognitive and evaluative framework that selects and structures their experience. It is well known that professions tend to conservatism in all ways; this is natural for prudent practice, on behalf both of clients and also of the profession. Formerly, when professions interacted only with their individual clients, such tensions could be managed outside the sphere of general politics. But now that issues of health, safety and the environment become ever more urgent, professional practice experiences strains analogous to those that the applied sciences have encountered in their confrontations with technological risks in recent decades (20). Thus the dialectic of the societal and environmental interactions of science has reached a new phase. It is now particularly confusing, since many scientists, and also their institutions, are only beginning to be aware even of the earlier change in the situation of applied science. But our present age is characterized by what has been called "future shock", where the very rate of change by itself creates problems of incomprehension.

5.6 Three types of problem-solving strategies

The inherent limitations of professional consultancy are revealed by a structural feature of the new global environmental issues. For in these, decisions depend on evaluations of future states of the natural environment, resources, and human society, all of which are unknown and unknowable. The powers of science have not only produced irremediable uncertainties in knowledge; now we also find moral uncertainties, resulting from the invasion of the domains of the sacred and private. The most notable cases here are reproductive technology and also scientific research that requires the inflicting of pain on aware beings. Under these circumstances of radical uncertainty, a new type of problem-solving strategy is emerging. In post-normal science, the traditional description, "the art of the soluble" is no longer appropriate. For in this work, it is issues rather than problems that are engaged upon, however much special scientific researches may be conducted and professional consultancy utilized. Instead of the traditional images of conquering or managing, now it is better to think of coping and ameliorating. This is a far cry from the old excitement of scientific discovery or engineering creativity; but now we must cope with the consequences of those traditional activities as they had been conducted for so long in innocence of their effects.

We can compare the different sorts of problem-solving strategies that are now employed, through a biaxial diagram which exhibits them in terms of the two attributes of "systems uncertainties" and "decision stakes", ranging from low to high, as on Figure 1 (21,22). For systems uncertainties, the three intervals along the axis correspond nicely with the distinctions we have already made among the different sorts of uncertainty, namely technical, methodological and epistemological. It is easy to see how the different types of practice correspond to these different sorts of uncertainty. The other axis of the diagram relates practice to the world of policy; and the zones on the two-dimensional provide a full specification for any issue. For decision stakes, we understand in general the costs, benefits, and commitments of any kind, for the various parties to an issue. There are three divisions, corresponding naturally to the three types of practice that we have discussed. In the case of applied science they are minimal; it is only exceptionally that a policy decision will depend on a single research result. For professional consultancy, they range from moderate to severe; the medical doctor normally cares for the health or life of a single patient, though he may also protect a wider community as with epidemiological problems; for the engineer there is the welfare of a client, and in connection with safety, that of a wider community. In post-normal science, when global environmental issues are involved, the stakes can become the survival of civilization as we know it or even of life on the planet. Although these distinctions are real, there is no pretence of quantifying either of the factors. The intervals, and the zones they define, provide a rough gauge which forms a part of an heuristic tool for distinguishing the three types of problem-solving strategies.

Looking now at the diagram, we see that applied science is performed when both factors are low; then puzzle- solving in the Kuhnian sense is adequate (23). But when either factor is medium, something extra must be brought into the work, which we can call consultant's skill or judgment. One very useful feature of the diagram is the way it displays the fact that even when uncertainties are low, if decision stakes are high then puzzle-solving alone will not be effective in a decision process. For no scientific argument can be logically conclusive; even the received views in the philosophy of science acknowledge this. Scientific arguments evolve in a continuous dialogue which is incapable of reduction to logic; what makes scientists 'rationally' change their opinions is a matter of ongoing debate among philosophers and sociologists (24).

Figure 1 Three types of problem-solving strategies

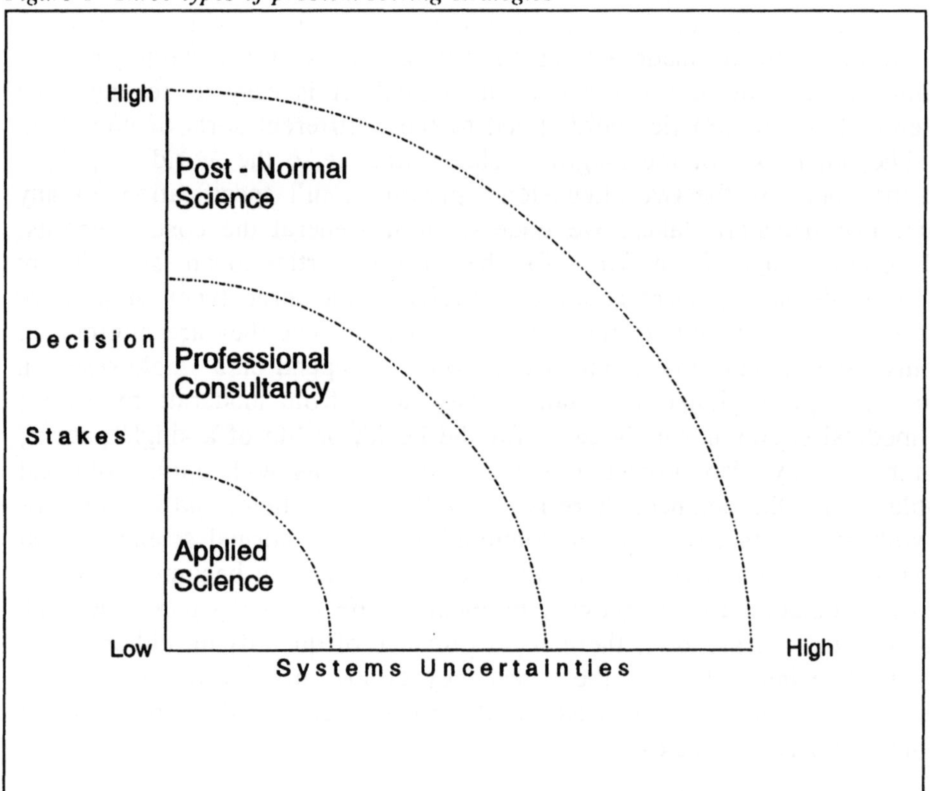

Applying this lesson to policy debates, we can appreciate that when a party finds its interests threatened it can always find some methodological issue on which to challenge results. This is particularly easy in the case of research on risks or the environment. Thus the forum for decision becomes enlarged from that of the technical experts, to include those with a strong stake in the outcome.

All these tendencies to debate appear still more strongly in the case of post-normal science. Although there is still an essential place for professional consultancy and even for applied science, the extremes of decision stakes or of systems uncertainties render them inadequate for the whole work. Research work and the deployment of skills have a central role to play, but this must be done in the epistemological framework in which the narrowly defined problems are integrated into larger issues. In this way they are provided with direction, quality assurance, and also the means for a consensual solution of policy problems in spite of their inherent uncertainties.

Examples of issues with combined high decision stakes and high systems uncertainties are familiar from the current crop of global environmental problems. Indeed, any of the problems of major technological hazards or large scale pollution belong here. The paradigm case for post-normal science could be the design of a repository for long-lived nuclear wastes, to be secure for the next ten thousand years. The strength of our diagrammatic scheme can be illustrated by consideration of cases located close to either of the axes. For a problem with low systems uncertainties, we have examples among the major disasters that have afflicted our modern industrial societies in recent years. Subsequent inquiries have in many cases established that the disaster had been 'waiting to happen' through a combination of physical predisposing causes and management practices which had been well known in advance (e.g. Bhopal, Challenger, The Herald of Free Enterprise, Exxon Valdez). Yet the processes of preventing a recurrence through improved regulations, or even of giving redress to the victims or punishing the culpable, can drag on for years or even decades.

A problem with low decision stakes will look very different; let us take for an example the field of cosmology. There the data are so sparse, theories so weakly testable, and public interest so lively, that the field is as much 'natural philosophy' as science; and experts must share the platform with amateurs, popularizers, philosophers and even theologians. In this latter example we see an historical continuity between the science that was practiced before the establishment of authoritarian paradigms, and the post-normal science of the present. This can help us appreciate the methodological continuity between applied science, professional consultancy and post-normal science. For post-normal science is a development from and extension of traditional science, appropriate to the conditions of the present age. Its essential principle is that uncertainty and ignorance, even in practice based on science, can no longer be expected to be conquered; instead they must be managed for the common good. Programs of reform of technology or lifestyle which ignore this aspect of knowledge are likely to remain part of the problem rather than contribute to its solution.

For the dominant historical experience within our present lifetimes is that science has created effective ignorance, in our inabilities to cope with the consequences of progress. Paradoxical as it may appear (and such apparent paradoxes may reveal the leading contradictions of an age) each advance in technique now opens up new areas of ignorance. These are not merely stimuli to curiosity-driven research, but they can threaten to vitiate the practice itself, unless they are appreciated as part of an enriched conception of knowledge. Merely to see ignorance as a negativity and a threat, is to remain in the old scientific paradigm. In the philosophy that we are now articulating, ignorance is a vital complementary aspect of knowledge, and in many cases becomes the driving force of progress. In this way, ignorance can perform the same functions for scientific methodology as the infinite for mathematics. By definition, the infinite cannot be known completely, but through its fruitful contradictions it has created, not only powerful mathematical tools and also beautiful structures, but even new conceptions of mathematics itself. In such a context we can genuinely speak of "usable ignorance", with the understanding that this is very different from "usable knowledge". For ignorance is usable when it is an object of awareness, and shows its dynamic interaction with knowledge.

By the use of the diagram, we can better understand the different aspects of complex projects in which all three sorts of practice may be involved. For this we may take an example of a dam, that was discussed previously (19) in connection with an analogous classification of problems as scientific, technical and practical. First, in the construction of a dam there is much basic, accepted scientific knowledge that is deployed; and there will be particular research projects of an 'applied science' character to determine the relevant features of the local environment for the dam and the details of its construction. But the making of the dam is in the first place a design exercise, where the shape and structure is not determined by the scientific inputs. If nothing else, there will be a design compromise among the various possible functions of the completed dam, which may include water retention, hydroelectric power, flood control, irrigation, and leisure, together with their associated costs. Achieving the optimum balance among these, given both the uncertainties in scientific inputs and the value-conflicts among interests, is a task for a professional. But the matter does not stop there. Some people may find their homes, farms and religious monuments drowned by the artificial lake; can they possibly be adequately recompensed? There may be a possibility of long-term deterioration of the hydrological cycle in the district, and perhaps even local earthquakes. Dams, once seen as a completely benign instrument of human control over raw Nature, have suddenly become seen as a sort of predatory centralism, practiced by vast impersonal bureaucracies against local communities and the natural environment. When such issues are in play, we are definitely beyond professional consultancy and in the realm of post-normal science.

We can also use the diagram to illustrate how a problem can evolve so that it is tamed, and brought some way in towards manageability. For when (for example) a risks or pollution problem is first announced, it will almost be in a condition of considerable uncertainty. Since it had not been appreciated previously, there is hardly likely to be substantial evidence about it. Hence the information will tend to be anecdotal on the experimental side and speculative on the theoretical side. But the strength of the decision stakes will ensure that all interests, aided by the independent media, will offer their opinions with apparently complete certainty. The first phase of the discussion will therefore resemble ordinary political debate, but of a particularly confused kind. For each side will attempt to define the problem in the terms most favourable to its interest, typically proponents presenting it as applied science and opponents stressing its uncertainties and also its ethical aspects. It is a new phenomenon for such debates to be effective; hitherto commercial viability or State security was the overriding consideration for industrial development, subject to a natural concern for health and safety. Indeed, in recent decades scientists and engineers have experienced bewilderment and dismay in confrontation with those who try to block progress on the basis of such intangible and non scientific arguments. One of the last debates of the old sort was that over

Recombinant-DNA research in the 1970's, when the evolving problem was kept firmly in the control of the scientists (25); in its sequel, on Genetic Engineering, the critics have scored some signal successes, as in Germany, and are now generally accepted as legitimate participants to the debate (26).

If such issues remained in the realm of pure power-politics, the outlook for our policies for science, technology and the environment would be grim. But there is a pattern of evolution of issues, with different leading strategies coming to prominence, which gives hope that the science may yet have an important role in such debates. For as the debate develops from its initial confused phase, positions are clarified and new research is stimulated. Although the definition of problems is (as we have seen) never free of politics, an open dialogue ensures that such considerations are neither one-sided nor covert. In the developing discussion on the technical aspects, no advocates need admit they were wrong; it is sufficient for there to be a tacit shifting in the terms of the dialogue. And as new research eventually brings in new facts, the issue becomes more amenable to the approach of professional consultancy. A good example of this pattern of evolution is Lead in petrol, where in spite of the absence of conclusive environmental or epidemiological information, a consensus was eventually reached that the hazards were not acceptable.

Thus the simple diagram of the three strategies for problem-solving enables us to see where traditional scientific practice is not effective, and why new dimensions need to be added to the problem-solving process. By its means we can make a dynamic analysis of the evolution of an issue involving science and policy. Post-normal science is thereby given its place as a complement to the other more traditional problem-solving strategies.

5.7 Quality assurance and post-normal science

It is important to appreciate that post-normal science functions as complementary to applied science and professional consultancy. It is not a challenge to the traditional practice of science, nor does it contest the claims to reliable knowledge or exclusive expertise that are made on behalf of science in its legitimate contexts. Recent critical philosophies of science, concentrating on scientific knowledge alienated from its social context, have led to a view that 'anything goes' in science. It is as if any charlatan and crank should have equal standing with qualified scientists or professionals (27). Our critical analysis proceeds on another basis, that of quality assurance. The technical expertise of qualified scientists and professionals in accepted spheres of work is not being contested; what can be questioned is the quality of that work, especially in respect of its environmental, societal and ethical aspects. Previously the ruling assumption was that these were somehow 'external' to the work of science itself; and that such problems as arose could be managed by some appropriate societal mechanism. Now the task is to see what sorts of changes in the practice of science, and in its institutions, will be entailed by these extension of the problems that are relevant to the quality of scientific and professional work. We have introduced these new aspects through the three strategies of problem-solving. Now we will develop their implications through an analysis of quality assurance in science.

Assessments of quality, and their use in quality assurance, have recently been appreciated as essential to successful practice in industrial production; this has been the lesson of the Japanese experience. We also know that many major disasters have been caused by defects, or low quality, at the interface between mechanical and information systems and the humans who operate and control them. Probabilistic analyses can indicate the relative likelihoods of different sorts of accidents, but real disasters tend to arise through sequences of events that no one had thought to put into the model. Now it is generally recognized that quality of information, as a component of real systems of communication, command and control, is critical.

As yet we are not so familiar with the idea of quality assurance in scientific information; yet it is equally fundamental. The public is familiar with the idea of scientific achievements of outstanding quality, as in the Nobel prize process. Those who award the prizes apply criteria of quality that are accepted for science; these include the extension of boundaries of the known, fruitfulness for further research, and also aesthetic considerations of elegance or surprise. Since only a small proportion of scientists receive such acclaim, the implication is that most results do not share such high quality. As in every other field of endeavour, most of the work is "average". Moreover, there will be work which is substandard in its quality, which may fail publication altogether or find a place in a less demanding publication. This principle is behind the grading of scientific worth through techniques of 'scientometrics', with the Science Citation Index as a well known research tool.

Such methods of grading for quality are widely used in the allocation of resources, for research projects and institutions. These frequently involve high decision stakes, in the sense we have used it here. However mathematical their form, these techniques are not an applied science; indeed, they provide a good example of how value presuppositions are built into the most apparently objective of research. Thus, the journals scanned by the Science Citation Index are necessarily a very small proportion of the total. By whom are they selected, and by what criteria and procedures? But then, who selects the selectors, etc.? The problem becomes acute in the case of evaluating science which is conducted outside the main metropolitan centres of research; in particular, the developing countries are systematically under-reported and under-rated through all the built-in biases of subject, choice of journal and institutions (28). This example also shows how a field which may claim to be one of professional consultancy actually involves such an interaction of decision stakes and systems uncertainties as to constitute a post-normal science.

Thus a formalization of the problem of quality assessment leads directly to an infinite regress in logic, and to informal, contested and confused practices in the actual work. In this way the assessment of quality in science, however 'objective' the products in question may be, shares the same problems as quality assessment in aesthetic productions; here the 'critics' may in the first place be scientists, but there is no science of selecting science-critics. Methodologically, this example shows the contradiction in the idea of a perfectly formalized system of knowing, where skills and judgements are to be excluded.

The idea of quality in science is foreign to the received philosophical view, where science is seen as "knowing-that", to the exclusion of "knowing-how", and is assumed to be solely about the eventual attainment of truth. The concepts of quality, and of controlled uncertainty, appropriate to knowing-how, have hardly any place in the traditional philosophies of science. These have been concerned with normative, idealized reconstructions of science rather than starting from the practice which has made science a model for successful human knowledge. This practice is of a specialized craft, whose subtle skills, including quality assessment, are passed down from master to pupil; in the absence of such a transmission of partly tacit information, quality of work inevitably degenerates.

If we keep the issues of quality in mind, we are in a position to understand how post-normal science is different from its predecessors. First, we recall that Kuhn defined "normal" science in his classic work The Structure of Scientific Revolutions (23). There he described it as "puzzle-solving" within an unquestioned "paradigm", an exemplar for practice. This is adopted by a subject-speciality community which consists of all those with the appropriate educational qualifications who also accept (generally unselfconsciously) common standards of quality on problems and on solutions. "Progress" takes place by means of such routine puzzle-solving, as the ruling paradigm becomes ever more articulated; indeed this is the defining property of normal science, in "matured" fields. Only when this approach fails to an embarrassing degree in resolving anomalies of practice, does the community lose its unanimity and undergo crisis. This leads to a "scientific revolution" and, eventually, the enthronement of a new paradigm, not so much building on the old one as replacing it and rendering all its associated puzzle-solving irrelevant and obsolete.

Kuhn did not merely describe the practice of 'mature' science; as an historian he was also very concerned to see how sciences achieved this state, out of the 'immature', embryonic early stages which they all went through. His account of the achievement of the state of 'maturity' has a most important ambiguity, which bears directly on the social aspects of post-normal science. For the transition he imagines a sort of 'social contract' among the practitioners of an immature field, in which they agree to a closure of the endless debate about foundational questions. Indeed, it was his experience of the contrast between argumentative behavioural scientists and acquiescent physical scientists that gave him the clue to the essentially social character of scientific maturity. He already knew that the foundational questions of physics are no more settled than those of psychology; why then do physicists normally ignore them? There seem to be two sorts of reasons: one is that puzzle-solving 'works' in a way that the community accepts as successful and progressive. The other is that dissidents are thenceforth ignored, or dismissed as nuisances. If we look at the unquestionably successful fields of academic scientific research, then the difference between these reasons is unimportant; only cranks and occasional rebels disagree with the consensus. But when we consider fields which are not so favoured, the ambiguity in Kuhn's picture becomes crucial. For we may then ask, how is the uniformity enforced? Kuhn himself indicated that there is something of a dogmatic, totalitarian element in normal science, as when he compared education in natural science to that in orthodox theology. This interpretation has been useful both to those academics who would like achieve maturity by fiat, forcing all members of a department to follow the line, and also to rebellious students who recognize no inherent virtue in the subject's ruling paradigm or its intellectual puzzles.

Kuhn did not deal explicitly with questions of quality in science, for his problem lay within the scope of the classic philosophy of science, concerned with claims to knowledge and progress. But his social model of scientific practice lends itself naturally to an analysis of quality assurance. Using his framework for the new problems facing science, we may say that in normal science quality assurance is effected by the closed community of practitioners on a well-defined set of problems on which they have exclusive esoteric expertise. In pre-normal science, quality assurance is a matter of continuous controversy, and this is taken as a sign of its immaturity. From the vantage point of post-normal science, we can appreciate the fertile ambiguity of Kuhn's formulation. For we are now familiar with the cases where a body of scientific or technological puzzle-solving is radically flawed or nearly vacuous when viewed from the outside, while the community of practitioners have by some means maintained a consensus that all is well or will be soon. It is difficult for a lay person to argue effectively that this or that field of academic science is not as mature as its proponents claim. But when the responsible experts are publicly unable to produce a class of environmental models that predict, or a class of space technology that performs, or an experimental practice that protects sentient beings, then by default there is an extension of the peer-community who exercise quality assurance. The reality as experienced by society at large then forces its way into domains that were previously the property of closed groups of experts.

The phenomenon of reality breaking into a social or intellectual system has recently been most obvious in connection with the societies of Eastern Europe, when the hollowness and ineptitude of the regimes' normal practice suddenly became a topic for public discussion and active dissent. Our technological systems have generally not suffered from such pathologies to the same degree within their defined spheres of operation and in their centres of origin; but the ongoing problems of both space and nuclear technologies, reminds us how competence is not to be taken for granted anywhere. We have previously described this as the 'Ch-Ch syndrome', after Challenger and Chernobyl (29). We could interpret the global environmental issues as an extension of this syndrome, where our scientific puzzle-solving has, all unknown to practitioners and general public alike, been seriously defective in important aspects of its quality.

Normal science still occupies a central position in the study of global environmental problems, in two ways. As part of the solution, there is always a need for scientific information which is as sound as it can be, much of it produced by well-tried techniques on limited problems. Indeed, as public interest in a problem leads to resources being invested in it, the relevant sciences gain in strength, and (as we discussed above) hopefully the problem is brought back in towards the 'professional consultancy' domain. In this way, the characteristic uncertainties are brought back from the epistemological (and perhaps ethical) state, towards the methodological and perhaps even technical. But normal science can also be part of the problem, in ways that are unfamiliar and disturbing to the individuals involved. For the global environmental issues have to a great extent been created by the practice of normal, puzzle-solving science and technology. Scientists and engineers who always thought that their work was purely beneficial to humanity, either directly or indirectly, now discover new problems thrown up by their past successes. Worse, their training and their inherited approach do not equip them for the solution of the problems directly associated with their work. Thus nuclear physicists are not skilled in oncology or epidemiology; nor are molecular biologists familiar with microbial ecology. Even more, specialists in human reproduction engineering are not systematically educated in ethics. Hence when such people in matured sciences try to cope with the problems created by their work, whether they manifest as hazards, pollution or ethical dilemmas, they are no longer working within their disciplinary paradigm. In solving such problems they are as amateurs, perhaps with a technical training that is very useful, but certainly not as puzzle-solvers within a secure scientific framework.

Traditionally trained scientists who venture into the fields of post-normal science thus find themselves in unfamiliar territory. The relevant disciplines (such as toxicology, epidemiology, ecology, and risk analysis) are weaker, technically and socially. They deal with more complex systems, are less well developed theoretically and historically have tended to lack prestige and resources. Furthermore, their relations with the public are very different. It is not a case of popularising esoteric results to an appreciative lay audience. Rather, the sciences address the worries of people, as residents, parents and human beings, perhaps even the families of those involved in creating the problem in the first place. The criteria of quality are broader than (say) theoretical interest or industrial applicability; they include considerations of health and well-being, of the environment and of humanity. The forums in which issues are debated are not restricted to the closed communities of subject-specialists, but will involve the media and various sorts of tribunals. In these, the scientist is not protected by his academic qualifications, and may be subjected to criticisms and interrogations which he may justifiably consider to be unscientific and unfair. In spite of these personal and professional hazards, the narrowly defined puzzle-solving disciplinary community can no longer maintain a monopoly on the quality assurance of their work; and so normal science must in these fields be superseded. There is a need for a new, more pluralistic strategy of inquiry, where the power embodied in quality assurance is more equitably shared among those with a legitimate concern for the consequences of scientific and professional work.

5.8 Post-normal science in historical perspective

Given all the continuities of history and method, there are disciplines involved in post-normal science which are radically different from those of the normal science which Kuhn took as his standard when analyzing "revolutions". In some respects they resemble the "immature" or "pre-paradigmatic" sciences out of which the traditional disciplines emerged. But it would be very misleading to describe them in that way, for the two terms have connotations of an incomplete growth towards a well-defined normal state. In the present case, we have scientific disciplines which cannot be expected to attain the normal state where routine puzzle-solving is effective for progress. Indeed, as we have seen in the case of computer models and environmental statistics, it is highly misleading to judge them by the criteria of quality appropriate for the traditional normal sciences. As scientific facts of the traditional sort, their quality is extremely low; but as inputs to a decision process where they serve as one sort of evidence among others, they have their genuine, indispensable uses.

It is useful to stress the historical context of these new sciences; this is why we call them "post-normal". For they are characteristic of an age when the old academic science, fractured by specialization, is becoming obsolete as the leading form of practice. Their immediate predecessor, and progenitor, is the "industrialized" science of the postwar period, when the capital-intensity, overall size and immediate applicability of scientific research made it truly, in J.D. Bernal's words, the "second derivative of production". Out of the laboratories of this industrialized science have come the techniques and processes which have done so much, first to bring comfort and convenience to so many of the world's inhabitants, and then to create the global environmental threats which we have recently come to appreciate.

There are many scientists and scientific advisors who still believe that these problems can be solved through the application of more normal science. In the terms of our analysis of problem-solving strategies, they cannot imagine anything other than "applied science" as being effective, for that is all that their philosophical formation and technical training allows for. For that reason it is particularly important to have a name that is easily remembered and that carries its meaning within it; and "post-normal", recalling the Kuhnian revolution in philosophy of science, is very appropriate in that regard.

Our concept of post-normal, like any other, has an intellectual ancestry. We recall that Kuhn himself was in rebellion against the prevailing positivist philosophy, which (partly unselfconsciously) promoted a certain ideological interpretation of science, as the unique bearer of the Good and the True. He carried the Popperian critique a stage further towards the direction of relativism; and where he feared to tread, Feyerabend and others rushed in (30). It is clear that within the terms of a narrow epistemological conception of scientific activity (however enriched by an analysis of social practice), there can be no solution of the sceptical crisis initiated in the turbulent 60's. Thus, in its own abstract way, the academic philosophy of science reflects the crises of confidence caused by the moral decline of science (seen in chemical, biological and nuclear armaments) and its practical impotence in the face of science-based environmental threats.

It is significant that the assessment of technological risks has motivated earlier initiatives in a reformation of scientific epistemology, along with our own. For here we face the paradox of an expertise whose form of argument resembles 'applied science', whose conclusion declare numerical measures of safety, which seems to be essential for any rational policy on management of risks, and yet which on examination reveals itself as shot through with uncertainty and subjectivity. In straightforward industrial practice it can be characterized as 'professional consultancy'; but for novel or complex hazards it is definitely a post-normal science. A pioneering effort to comprehend this new form of practice was made by A. Weinberg, when he created the concept 'trans-science', in connection with hazards of low-level radiation. On examination, this could be seen to be saying both too little and too much. On the one hand, the name itself implied that here was something that was not science as we understand it, but a practice that is essentially different. But in reply to critics, Weinberg admitted that his only demarcation criterion was the one of scale, that is the feasibility of a project in relation to existing resources as willed by society. Given more resources or 'more sophisticated science', a trans-scientific problem can be rescued for science (31). Thus 'trans-science' turned out to be a distinction of quantity only, while our concept of post-normal science, involving new methods and new societal practices, represents a qualitative transformation of science.

Nearly ten years ago the philosopher of science Stephen Toulmin suggested the term "post-modern" for this new science as it is actually practiced. Modern science, descended from Galileo and Descartes, involves an alienation of the scientist, the atomization of knowledge, and the neglect of systems as wholes. Toulmin called for recognition of scientists as actors, and of socio-ecological systems as wholes (32). Further, it has been argued that this new sort of science must follow Gregory Bateson's teaching (33) and be at least as much concerned with information, meaning and motive as with measurements and physical causation (34). This valuable initiative was too ambitious in some ways and insufficiently so in others. The term post-modern has a very wide currency, and a family of meanings which have generated a considerable academic industry. So much of its programme involves the denial of the reality of the traditional epistemological issues, that it is hard to see how it can be related to any practical concerns where the quality and reliability of science are important.

Another strand of research complementary to our own, is that of the "cultural theory" school. They study issues with "structural" uncertainties as distinct from merely "technical". This corresponds to our post-normal science, though without the fruitful analysis of "systems uncertainties" and "decision stakes". But they provide an enriched framework for the perceptions of the various participants in such issues, including "hierarchists", "individualists", "egalitarians" and "fatalists". On that basis they can offer suggestions for political processes whereby issues in post-normal science can be resolved (35).

Our work on policy-related research began with the very practical problem of the public debates on "acceptability of risks", where the mathematical form of the experts' argument did not guarantee the acquiescence of critics, but instead only fuelled further methodological debate. Could a more refined mathematical formalism, encompassing more vague and ambiguous judgements, be devised so that the Leibnitzian programme of calculating solutions to all arguments could be fulfilled in this case? We doubted this; for such programmes had already failed in far less difficult fields of discourse. We decided to treat uncertainty with respect rather than attempting to banish it with formulae; accordingly we proceeded to analyze the different sorts of uncertainty that affect all quantitative statements. We found ourselves moving out into very new ground, as our philosophy of mathematical knowledge developed in a dialectical and pragmatic direction. But we were always kept secure against the pitfalls of relativism by our concern with the resolution of real problems of practice, and by our commitment to the enhancing of the craft skills of those who use mathematics in these new and confusing situations. From our analysis of uncertainty we moved on to that of quality, and from there we arrived at the issues of who is competent to assess quality; and this was in the context of risk assessment with which our work began. The centrality of the concept of quality to a new epistemology of science had already been realized in earlier work (19); in this present work that insight was articulated and given a practical focus. Hence our conclusions about the new social practice of science are firmly based on our prior epistemological studies, themselves conducted in a dialogue with the main stream of contemporary philosophy of science (36).

5.9 Social aspects of post-normal science

In what we might now call "pre-normal" science, nearly all the practitioners were amateurs. They could and did debate vigorously on all aspects of the work, from data to methodology, but there was no in-group of established practitioners in conflict with an out-group of critics. In normal science, any outsiders were effectively excluded from dialogue; only in a Kuhnian "pre-revolutionary" situation, when the ruling paradigm (cognitive and social) could not deliver the goods in steady progress, would outsiders get the chance to be heard. In post-normal science there is still a distinction between insiders and outsiders, based (on the side of knowledge) on certified expertise and (on the social side) by occupation. But since the insiders are manifestly incapable of providing effective conclusive answers to many of the problems they confront, the outsiders are capable of forcing their way into a dialogue. When the debate is conducted before a lay public, the outsiders (including community activists, lawyers, legislators and journalists) may on occasion even set the agenda.

Because of these human aspects of the issues giving rise to rise to post-normal science, there must be an extension of all the elements of the scientific enterprise. First there must be a presence of an expertise whose roots and affiliations lie outside that of those involved in creating or officially regulating the issue. These new participants, enriching the traditional peer communities and creating what might be called "extended peer communities" are necessary for the transmission of skills and for quality assurance of results. For in the case of the new sort of science, who are the "peers"? In Kuhn's normal science, they are colleagues on the job, engaged in that "strenuous and devoted effort to force Nature into the conceptual boxes provided by professional education." Such peers are still there, as scientists and experts; and they exercise quality control within the technical paradigm of their expertise. But the problems of the new sort of science are not ones of purely knowing-that within stable paradigms; they include knowing-how, along with broad and complex issues of environment, society and ethics. Hence it is necessary and appropriate for quality assurance in these cases to be enriched at the very least by the contribution of other scientists and experts, technically competent but representing interests outside the social paradigm of the official expertise.

It is important to realise that this phenomenon is not merely the result of the external political pressures on science that occur when the general public is concerned about some issue. Rather, in the conditions of post-normal science, the essential function of quality assurance can no longer be performed by a restricted corps of insiders. When problems do not have neat solutions, when the phenomena themselves are ambiguous, when all mathematical techniques are open to methodological criticism, then the debates on quality are not enhanced by the exclusion of all but the academic or official experts. Knowledge of local conditions may not merely shape the policy problems, but it can also determine which data is strong and relevant. Such knowledge cannot be the exclusive property of experts whose training and employment inclines them to abstract, generalized conceptions. Those whose lives and livelihood depend on the solution of the problems will have a keen awareness of how general principles are realized in their "back yards". It may be argued that they lack theoretical knowledge and are biased by self-interest; but it can equally well be argued that the experts lack practical knowledge and have their own forms of bias.

An appreciative study of local knowledge in solving scientific and technological problems is only now getting underway. Some authors have recognized this as the key to genuinely sustainable development. The author Arnold Pacey gives examples to show how a really successful technology is the outcome of a "dialogue" between what is an apparently more advanced innovative culture, and the apparently traditionalist receiving culture. Thus in African agriculture, the previous dominance of colonially-introduced temperate-zone concepts is being replaced by the integration of tree and field crops (incomprehensible to Western experts), together with irrigation and minimal engineering (37).

In Europe, a recent survey by Brian Wynne of the University of Lancaster has shown how the sheep farmers of Cumbria in England have a better understanding of the ecology of radioactive deposition than the official scientists (38). The farmers would not have made the assumption that radioactive contaminants would drain away through their thin cover of moorland soil at the same rapid rate as through lowland pastures. Also, they would have recognized that high ground lying directly downwind of a major reprocessing plant (the nearby Sellafield plant of British Nuclear Fuels Ltd.) is liable to have a different deposition pattern from remote fields. Although they could not criticize the technically esoteric measurements made by the official scientists, they were fully competent to evaluate their methods and interpretations at every stage.

Along with the enrichment of the traditional scientific peer-communities we have a parallel enrichment of the cognitive basis of post-normal science; we speak of "extended facts". This is the material which is effectively introduced into a scientific debate on policy issues. It is now widely appreciated that the beliefs and feelings of local people, whatever their source and validity, must be recognized and respected lest they become totally alienated and mistrustful. But extended facts go beyond that purely subjective base. There will also be anecdotes circulated verbally, and then the edited collections of such materials prepared for public use by citizens' groups and the media. These will not usually be of traditional scientific form, but they may be essential for establishing a prima facie case for the existence of a problem, and therefore the urgency of systematic research. When such testimonies are introduced into scientific debate, and subject to some degree of peer-review before reporting or acceptance, they approach the status of scientific facts. Of similar strength are the experiences of persons with a deep knowledge of a particular environment and its problems, like the hill farmers of Cumbria reported by Wynne. We should not forget material discovered by investigative journalism. Finally, the category of extended facts can also be applied to information which is quite orthodox in its production, but which for political or bureaucratic reasons is officially secret in some way or other; they can then function covertly, forming a background to loaded public questions. This last sort of "fact" may seem very strange to those whose idea of science is derived from the textbook and the academic research laboratory. But for those who are familiar with science in the policy context, such extended facts may be quite crucial in the accomplishment of the quality assurance of results on which health and safety depend.

As post-normal science depends so critically on data which are frequently inadequate in quantity and quality, the pitfalls in its production and interpretation are particularly severe. Scientists who are engaged on an academic exercise, or those working for a bureaucracy with a vested interest in the issue, will not normally be inclined to check for all the possible hidden traps that could vitiate their results. It is entirely natural and appropriate for those with a personal interest in the issue, and a personal knowledge of the phenomena, to engage in a dialogue on quality assurance. As yet this has happened only sporadically, and in a context of conflict and polarization of interests. The task is to create the conceptual structures, along with the political institutions, whereby there may be developed a creative dialogue. For this, post-normal science is a foundational element.

5.10 Philosophical perspective on post-normal science

In the title of this essay we introduced two terms; one, quite familiar, set the problem; the other, a neologism, announced the path to a solution. Our argument is that the novel challenges of global environmental issues will not be met by the old strategies for scientific problem-solving. As the social practice of science has evolved through the ages, it is now ready for another mutation; one required by the new circumstances of humanity, but at the same time made possible by its new material and cultural conditions. We have seen that the existing techniques for forecasting about the environment cannot produce data in which uncertainty is sufficiently reduced for effective predictions to be possible. Nor indeed can the approaches of applied science, or even professional consultancy, be adequate in the many situations in which systems uncertainty and decision stakes are high. In thinking about science in its social and natural environment, we now confront our ignorance, as never before. It appears in the form of contradictions, as that our science-based technology produces environmental problems which may well be beyond the capacity of science-based technology to solve. Faced with such unprecedented challenges to our reason, some may well despair of science, and retreat to some otherworldly inspiration for support and guidance. Our programme is not to abandon or condemn science, but to find a way to enrich it, to foster a rejuvenated science which would help in the transformation of knowledge as a human possession along with the transformation of technology in its environmental context.

The title "post-normal" announces a break with a previous state; the "normal" practice to which it refers is the puzzle-solving research, conducted within subject-specialty communities that are alienated from their societal and natural environments. There is no question that this practice had succeeded brilliantly in its own terms, and in its by-products of technological advance, for generations. But it now experience crises of ever increasing severity, and in the terms of the theory in which it was originally cast, it will soon be ready for a "revolution". This will not be within this or that field, but, rather like the original "scientific revolution" of the seventeenth century, will affect the definition of the objects, methods and social functions of science itself.

Our path to this new definition is laid out in the diagram of the three strategies for problem-solving. There is a methodological continuity as well as an historical continuity, between the three forms of practice. We do not believe in the possibility of revolutions in which the past is merely discarded and not included while being transcended. Given all the contradictory relations of applied science and professional consultancy with post-normal science (for as we have seen they can also be part of the problem as much as of the solution), without reliable knowledge and certified skills there can be no hope of solving the great global problems.

The dialectical relations of post-normal science with its constituent traditions extends further back, indeed much further back than the current century's attempt to rescue the Good and the True in science. The problem of knowledge, focussed on scientific knowledge, has an intellectual continuity back to classical Greek times. We may identify three sorts of themes, interweaving in philosophical thought between then and now. In the middle, as it were, is scientific knowledge, using systematic demonstration and based on reason and sense-experience; this was defined and given its first examples by Aristotle. To one side lay practical, craft experience, realized either in the liberal arts of persuasion, or in the manual arts of working on matter. To the other side lay inner experience, approached either through wisdom or through enthusiasm. The interactions between these three themes down through history are not our present concern; let it suffice that since the Scientific Revolution, pure scientific knowledge has been prized above the others, and it has been claimed on its behalf that all their proper claims to knowing could be realized through it.

Now that the certainties of science have been betrayed by its very successes, we have the occasion to reconsider our own place in the interweaving of these three themes. The third theme is beyond our present concerns; our focus is on the world of shared experience and how humanity can find its true place in it. But the second theme, that of practice, is due for a radical re-evaluation; and our analysis of the problems and of their solution through post-normal science, may show the way to that. For that, we consider the two aspects of the theme of practical knowledge. The first, the rhetorical tradition, has had a very mixed reputation ever since Plato attacked the its practitioners for their lack of concern with truth; but it has been a recurrent minor theme in philosophy ever since. The second, the tradition of manual arts, has inevitably suffered from its association with people who lacked the capacity for argument and reflection. One of the most important things about the Scientific Revolution was its re-evaluation of the manual arts, so that a philosophy of practice became a part, however briefly, of a philosophy of knowledge. Afterwards, leading philosophers were either indifferent or hostile to the claims of either of these strands in the practical tradition; and the problem of knowledge conceived narrowly in Aristotle's terms has dominated intellectual effort.

We have now reached the point where that narrow tradition is no longer appropriate to our needs. Unless we find a way of enriching our science to include practice, we will fail to create methods for coping with the environmental challenges, in all their complexity, variability and uncertainty. Fortunately, the conditions are ripe, in the changing social distribution of knowledge and skills. For now the liberal arts, as rhetoric, are no longer restricted to a tiny privileged elite in society, and the manual arts have lost the stigma of belonging to the oppressed majority. The improvement of manners and morals, from the Enlightenment through industrialized society, has been real. In modern societies there are now large constituencies of ordinary people who can read, write, vote and debate. The democratization of political life is now a commonplace; its hazards are accepted as a small price to pay. Now it becomes possible to achieve a parallel democratization of knowledge, not merely in mass education but in enhanced participation in decision-making for common problems.

The democratization of science in this respect is therefore not a matter of benevolence by the established groups, but (as in the sphere of politics) the creation of a system which in spite of its inefficiencies is the most effective means for avoiding the disasters that in our environmental affairs as much as in society, result from the prolonged stifling of criticism. Let us be quite clear on this; we are not calling for the democratization of science out of some generalized wish for the greatest possible extension of democracy in society. The epistemological analysis of post-normal science, rooted in the practical tasks of quality assurance, shows that such an extension of peer-communities, with the corresponding extension of facts, is necessary for the effectiveness of this new sort of science in meeting the challenges of global environmental issues.

5.11 Political epistemology and post-normal science

By now it is clear that our enterprise is not one of traditional philosophy of science as it has been practiced in recent generations. Our concern throughout is with practice, the sort of science that is necessary for meeting the challenges confronting our civilization. In that sense it is itself an example of the 'post-normal' phenomenon. Yet we are doing philosophy, engaging in a conscious dialogue with those whose insights we use while criticizing them. We are attempting to see how our new formulation of the philosophical task can take us through the barriers that prevented them from ever speaking coherently to the issues of our age. For this reason we review the doctrines of our predecessors, gaining clarity on how their conception of the problem constrained their solutions in ways that could not be surmounted.

In proposing a new form of science, we are re-defining knowledge; and that is the task of the branch of philosophy called epistemology. Our re-definition is in the direction of taking knowledge out of the classroom and the laboratory, into the broader community of people in their man-made and natural environments. In this sense our epistemology is political, not dealing with party-politics in the ordinary sense, but with what the Greeks called the polis (39,40). We could also call it ecological, from the Greek word oikos for household, from which both eco-nomics and eco-logy are directly derived. Indeed, we could say that our political epistemology is an oikos-philosophy; where our whole earth is a household. Such an image, midway between the technocratic 'spaceship earth' and the goddess earth 'Gaia', seems best suited for the conception we want to develop.

Our dialogue with previous epistemology goes back to Aristotle, whose ideal of science was fundamentally deductive, on the example of geometrical argument leading to truth. The complementary aspect was stressed by Bacon, who believed in induction from particular experiences as the way to truth. Only in the present century was the ideal of truth as the goal of science become attenuated; with Popper the task was the demarcation of science from its imitations, and falsifiability became the criterion; truth was left as the distant goal of theories with increasing empirical content. This line could not be held, and Kuhn offered the alternation of dogmatic "normal science" with irrational "scientific revolutions, with a product that promised neither truth nor even progress in the long run. Finally, the end of classical epistemology came with Feyerabend and his Dada-science; his counter-method is 'anything goes', and its outcome is that nothing happens after Woodstock.

This sequence of retreats and betrayals of reason in recent decades has its own internal logic. For so long as the problem is cast in abstracted terms of the achievement of truth (necessarily seen as some sort of absolute), then as soon as criticism is heeded there is no lasting defence of a position that is fundamentally brittle. Among philosophers the task of epistemology has commonly been seen as the refutation of 'the sceptic';

and with the requirement of 100% security for success, it is no surprise that failure has been the outcome. Our approach has been to appreciate Truth as a regulative principle for our striving, but one which can no more be pinned down than any of the other absolutes as the Good or the Beautiful. One of the keys to post-normal science is the insight that the Safe belongs to the same class; although everyone wants it, it cannot be calculated or guaranteed by any routine methods, of applied science or professional consultancy.

Our model for scientific knowledge is not derived from its rational reconstruction, but from its actual practice as revealed in historical, reflective and critical studies. We see it as the synthesis of several complementary polarities: theory/practice, knowing-that/knowing-how, facts/values and knowledge/ignorance. For us knowledge does not advance or grow by simple accumulation along secure pathways, but it becomes alive by the force of conflict and contradiction. Its objectivity is achieved not through logic, but by a social process of the application of craft skills, guided by ethical principles. By its style it is resistant to the processes of reduction to atomized academic specialties, while yet keeping its core of genuine expertise intact.

The philosophical core of our programme is the twinned concepts of uncertainty and quality. With them we find the bridge between the subjective and the objective, the epistemological and the axiological. When we recognize inescapable uncertainty in genuine scientific knowledge, we know that it is not reducible to the caricature of True/False. As we have seen, it possesses many sorts of uncertainty, all of them related to quality. Quality is relative to function, but it is not arbitrary on that account. Although a 'pragmatic' relation, it is not thereby reduced to a trivial or superficial evaluation. Quality is used in many distinct senses in all spheres of practice, and is quickly comprehended as a basic idea which has hitherto been neglected. With quality we can renew our engagement with the perennial questions of philosophy, hopefully on this occasion with an approach that will be fruitful for theory and practice alike. With uncertainty and quality as the central concepts for the science of the present age, we have seen how extended peer-communities and extended facts are natural and inevitable enrichments of previous scientific practice.

Notes

1. Reprinted with permission from C. Rossi and E. Tiezzi (eds.) *Ecological Physical Chemistry, Proceedings of an international Workshop*. Copyright 1991, Elsevier Science Publishers B.V.

References

1. S. Mac Lane, *Science*, 241 (1988) 1144.
2. N. Keyfitz, *Science*, 242 (1988) 496.
3. S. Mac Lane, *Science*, 242 (1988) 1623 - 1624.
4. J.C. Bailar, *Scientific Inferences and Environmental Problems: The uses of Statistical Thinking*, Institute for Environmental Studies, The University of North Carolina, Chapel Hill, 1988.
5. J.R. Ravetz in: (41) 260-283.
6. M. Smithson, *Ignorance and Uncertainty*, Springer-Verlag, New York, 1989.
7. W.C. Clark and G. Majone, *Science, Technology and Human Values*, 10(1985) 6-19.
8. J.V. Rivard et al, NUREG/CR-3440, SAND 83-1689, USNRC, Washington, 1984.
9. M.C.G. Hall, in: M.C. MacCraken and F.M. Luther (eds.), *The Potential Climate Effects of Increasing Carbon Dioxide*, DOE/ER-0237, USDOE, Washington, 1985, 337-364.
10. M.B. Beck, *Water Resources Research*, 23 (1987) 1393-1442.
11. HSE, Canvey: *An Investigation of Potential Hazards from Operations in the Canvey Island/Thurrock area*, London, 1978, 48.
12. UNEP, GC.12/11/Add.2, 1984.
13. British Standards Institution, BS 4778, London, 1979.
14. M. Thompson and M. Warburton, *J. Applied Systems Analysis*, 12(1985) 3-34.
15. S.M. Macgill, *The Politics of Anxiety*, Pion, London, 1987, 141-156.
16. S.O. Funtowicz and J.R. Ravetz, *Uncertainty and Quality in Science for Policy*, Kluwer, Dordrecht, 1990.
17. US NRC, NUREG- 75/014, WASH 1400, Washington, 1975.
18. M. Polanyi, *Personal Knowledge*, Routledge and Kegan Paul, London, 1958.
19. J.R. Ravetz, *Scientific Knowledge and its Social Problems*, The Clarendon Press, Oxford, 1971.
20. D.A. Schon, *The Reflective Practitioner*, Basic Books, New York, 1983.

21. S.O. Funtowicz and J.R. Ravetz in: C. Whipple and V. Covello (eds.), *Risk Analysis in the Private Sector,* Plenum, New York, 1985, 217-231.
22. S. Rayner, S in: H. Jungermann, R.E. Kasperson and P.M. Wiedemann (eds.), *Risk Communication,* KFA, Julich, 1988, 169-176.
23. T.S. Kuhn, *The Structure of Scientific Revolutions,* University of Chicago, 1962.
24. A. Chalmers, *Science and its Fabrications,* Open University, Milton Keynes, 1990.
25. J.R. Ravetz in: (41) 63-80.
26. J.R.S. Fincham and J.R. Ravetz, *Risks and Benefits of Genetically Engineered Organisms,* Open University, Milton Keynes, in press.
27. P.K. Feyerabend, *Against Method,* New Left Books, London, 1975.
28. M.J. Moravsik, *Scientometrics,* 7(1985) 165-176.
29. J.R. Ravetz, S. Macgill and S.O. Funtowicz, Disasters Bring the Technological Wizards to Heel, *The Guardian,* London, 19 May 1986, 22.
30. J.R. Ravetz in: (41)180-198.
31. A.M. Weinberg, *Science,* 180(1973) 1124.
32. S. Toulmin, *The Return to Cosmology,* University of California, Berkeley, 1982 .
33. G. Bateson, *Steps to an Ecology of Mind,* Ballantine Books, New York, 1972.
34. R. Rappaport, *Risk Analysis,* 8, 2(1988) 189-191.
35. M. Swarz and M. Thompson, *Divided We Stand: Redifining Politics, Technology and Social Choice,* Harvester Wheatsheat, London, 1990.
36. S.O. Funtowicz and J.R. Ravetz, *Global Environmental Issues and the Emergence of Second Order Science,* EUR 12803 EN, CEC, Luxembourg, 1990.
37. A. Pacey, *Technology in World Civilization,* Blackwell, Oxford and Cambridge MA, 1990, 203.
38. B. Wynne, Personal Communication.
39. M. Cini, Personal Communication.
40. M. O'Connor, *Time and Environment,* Department of Economics, University of Auckland, New Zealand, 1990.
41. J.R. Ravetz, *The Merger of Knowledge with Power,* Cassell, London, 1990.

PART III

The Use of Social Science in the Policy Making Process

6 ARGUMENTATION AND POWER IN EVALUATION-RESEARCH AND IN ITS UTILIZATION IN THE POLICY-MAKING PROCESS[1]

Igno M.A.M. Pröpper, Free University of Amsterdam

6.1 Introduction

> "I learned quickly that there was a vast difference between political and academic debate; any attempt to force the academic model on a campaign was, at best naïve, and at worst politically dangerous (Martel, debate adviser for American politicians, 1983, p. xi)".

The idea that politics is mainly a matter of power and manipulation is quite common. Names which spring to mind are: Machiavelli, Weber and Trotsky.[2] To those sharing this view, appealing to rationality is, in fact, an a-political method. This is confirmed by Hoogerwerf's research, in which he quotes a cabinet minister making the following statement:

> "Thinking logically and reasoning soundly are soon unlearnt in politics. A carefully balanced argumentation in the House is less understood than slogans" (Hoogerwerf, 1986, p. 271).[3]

The opposite view is less popular, namely that, in essence, politics is a matter of co-operation and information and therefore a matter of sound argumentation. There are people, like Bertrand de Jouvenel who believes that creating, enforcing and upholding channels of human co-operation is the purest form of political activity (quoted by Hoogerwerf, 1979, p. 41).[4]

If we turn to scientific discussion it would be fair to say that the scientific code requires them to proceed rationally.[5] Scientific knowledge is only valid when it can win general approval on the basis of sound arguments. Seeking after truth (or, in the case of normative scientific statements, justice) is pointless if forms of power, such as coercion, bribery or deception are used.

The above characterisation of scientific and political discussions raises the question as to what part scientific research might play in politics. Banner has the following to say in this connection, namely that evaluation research may become an instrument of political power:

"Unfortunately, because of the political environment surroundig evaluation research, 'objective' research often proves impossible. Evaluation can become a tool for wielding power and, as such, is constantly an active variable in changing power relationships within political and organizational structure". And:
..even the most carefully designed well-implemented evaluation research is often sabotaged by factors within the program's political environment (Banner, 1975, p. xv and p. 2)."

On this same subject Cronbach puts forward a normative point of view:

"Evaluations should contribute to wiser social actions (Cronbach, 1980, p. 6)".

In this chapter an account will be given of empirical research investigating the question of to what extent evaluation research results in decisions being taken in political discussions on the basis of the soundest possible arguments rather than on the basis of exercising power (section 8).

Before attempting to answer this question, it is important to determine the nature of the measuring instrument. When can we speak of sound political argumentation and when does power start to play a role? A large part of this chapter will be dedicated to the plan for such a measuring instrument, for use in empirical research into political - and also other - discussions.

First of all, the concepts of sound argumentation and power will be explained (section 2 and 3). Subsequently the question of the relation between sound argumentation and power in a political discussion will be defined and specified on the basis of this definition (section 4). Next, a number of argumentation rules for political (and other) discussions will be presented. On the basis of these rules, one can first assess the quality of a line of argumentation (section 5). Secondly, these rules help to establish how far power is being exercised in a political discussion (The latter will only be found in section 7). Next follows a strategy for the operationalisation of these rules for political discussions (section 6). Finally, in section 7, an answer can be given to the question of how, in a concrete political discussion, one is to determine to what extent sound argumentation plays a part and to what extent power is exercised.

6.2 Sound argumentation: a definition

When designing a measuring instrument, it is important to be quite clear about what is to be measured. For this reason the first two sections will provide a description of what is to be understood by the concepts of 'sound argumentation' and 'power'. At the same time, a number of other terms such as communicative (rational) and strategic acting will be introduced as well.

Let us start with the concept of *sound argumentation* and first look at the term 'argumentation'. Van Eemeren and Grootendorst describe *argumentation* as follows:

> "Argumentation is a speech act consisting of a constellation of statements designed to justify or refute an expressed opinion and calulated in a regimented discussion to convince a rational judge of a particular standpoint in respect of the acceptability or unacceptability of that expressed opinion (Van Eemeren & Grootendorst, 1983, p. 18)."

The question of to what extent an argumentation is *sound* will be the next consideration after dealing with the rationality of an argumentation. A large number of meanings is ascribed to the concept of *rationality* (see e.g. Dunn, 1981, p. 225; Rapoport, 1980). It is not within the scope of this chapter to go into this in detail. The starting-point will be a concept of rationality in which argumentation plays a major part, viz. the concept of *communicative rationality*, as developed by Jürgen Habermas (1981). Habermas grounds this concept on the acting type known as *communicative acting*. A person acts communicatively when he tries to come to an agreement with others on the basis of the soundest possible argumentation. The more successful one is, the greater the communicative rationality of an act will be.

For our purpose, the concept of communicative rationality is too wide to be used in this form. In order to make this clear, it is essential in the first place to differentiate between two aspects of the concept of communicative rationality viz. a material and a procedural aspect. Then an explanation will be given as to why we will restrict ourselves to the second aspect. The concept of communicative rationality allows an argumentation to be assessed both materially and procedurally. In the case of a *material assessment*, an opinion is given on the acceptability of the expression of an opinion to be justified or on the unacceptability of the expression of an opinion to be refuted. This is illustrated in the following example of an argumentation:

> The speed limit for motor traffic should not be increased to 120 km/h, because this is detrimental to the environment. For in case of a speed of 120 km/h more petrol is used than with the current speed limit of 100 km/h.

When materially assessing this argumentation the question is, to what extent the opinion that the speed limit for motor traffic should not be increased, is acceptable on the basis of the argumentation opted for. In the case of a *procedural assessment*, it is not so much the acceptability of an expression of opinion that is at issue but the acceptability of the style of reasoning. The correct way of acting in general can be judged by social rules,[6] whereas in particular cases the style of reasoning might be assessed on the basis of discussion rules. Examples of these rules are that one should not contradict oneself and that one adopts an objective attitude - the latter implies, among other things, that one is not allowed to make tendentious statements. In section 5, a number of rules for (political) discussions will be formulated in more detail. These rules will fall within the boundaries of the concept of communicative rationality. *For the time being*, we will refer to them as 'discussion rules' without going into further detail.

It might be argued that an approved argumentation procedure is an essential, but not a sufficient condition for the acceptability of the act or the opinion to be defended. Looking again at the above example, nothing seems to be wrong with the argumentation procedure, that is, with the style of reasoning. Nevertheless, the expression of opinion to be justified - that the speed limit should not be increased - isnot necessarily acceptable to everyone. One might base one's opinion that the speed limit should be increased on other considerations, e.g. because it would cut travelling time down to a minimum.

Summing up, on the basis of the concept of communicative rationality, an argumentation may be assessed in two ways. The following applies here:

1. The more the expression of opinion to be jusitified is acceptable on the basis of one's arguments, the greater the *material rationality*.
2. The more an argumentation is structured according to discussion rules, the greater the *procedural rationality*.[7]

If a researcher wishes to give a somewhat systematic and objective judgment about the presence or absence of a certain quality in an argumentation, materially speaking, he must be able to assess the acceptability of every expression of opinion on the basis of standards that can claim to be generally valid. If we only worked on the example given earlier of an argumentation about the speed limit for motor traffic one might conclude that we do not have such standards. If this were the case, most political, and scientific discussions would be superfluous. Therefore the quality of an argumentation will not be assessed materially, but procedurally. For, in our opinion, developing a general standard is feasible for the assessment of the procedural rationality of an argumentation. As already mentioned such standards can be drafted with the help of discussion rules. In order to prevent misunderstandings it should be noted even at this early stage, that establishing whether or not a discussion rule has been violated in a concrete case, does require a material assessment. However, to establish this empirically, no other standards are needed but the very same discussion rules. These rules will be dealt with in section 5.

6.3 Power: a definition

According to Ellemers, in most definitions *power* is described as follows:

(1) "Power is the capability or means to impose one's will on others, even when they do not want it."[8]

For theoretical reasons we consider this definition of power both too wide and too narrow. This will be explained, in brief, and the concept of power will take on a new meaning, only one of its aspects - to be called 'exercise of power' - will be dealt with in more detail. This is for reasons of a measurement-technical nature. Let us start with the theoretical considerations. First we shall indicate why the above description of power is too wide and has to be curtailed.

The main point of our study is to enable us to differentiate between power and sound argumentation in a concrete empirical situation. For this, it is essential that power and sound argumentation exclude one another, remaining separate concepts. In other words the two concepts must not overlap. This is, however, the case in Ellemers' definition of power. In order to make this clear, we will give a more precise definition:

(2) "Power is the capability or the means to impose one's will on others, whether they want it or not".
(Comparing the two definitions then "even when they do not want it" in the first description has been stated precisely as "whether they want it or not").

It appears that in the latter definition of power, it is also possible to impose one's will on another person when this person agrees with it. Apart from a linguistic objection - *imposing* something on someone in fact implies that one acts against another person's will - there is another objection. For according to this definition it is also a matter of power when a person conforms to another person's will by changing his initial opinion - of his own free will - on the basis of the other person's arguments. In the same strain Van Doorn and Lammers also speak of arguments as 'means of power' (Van Doorn and Lammers, 1976, p. 79/80). Here we observe that concept-wise argumentation and power partially overlap, and because this is just what we wish to preclude, the concept of power will be curtailed.[9] This is possible by considering the influence of arguments not as a form of power but as *authority* or put more precisely, as *communicative rational authority*.[10] We describe this as follows:

(3) A person has communicative rational authority when one or more other persons conform to his will on the basis of arguments.[11]

Having excluded this meaning from the concept of power, we get the following definition of power:

(4) Power is the capability or means to impose one's will on others.

As noted earlier in the discussion, linguistically, the term *'imposing his will on others'* is used somewhat carelessly in the literature. In order to do justice to the description of power above (4), this term will be described in more detail as follows:

A person imposes his will on another person, when the latter conforms to this will by doing or not doing something, without approval on the basis of a true conviction that what is to be done or not done is correct by the nature of things.

Summing up on the above, first we affirmed that our initial description of power (1) was too wide, upon which we formulated the altered version (4).

However, for quite different reasons, we consider both definitions to be narrow from another point of view. They only look at one aspect of power, without taking another side of power into consideration. To be quite candid, this one-sidedness occurs in every discussion on the concept of power in the literature. For with the concept of power, it is usually only the following model which is taken as a starting-point:

(I) An *actor a* is capable of making or trying to make another *actor b* do or refrain from doing something (by some form of coercion).

In this model actions of actor b are the result of the power of actor a. There is yet another form of power. In this form it is not so much the actions of actor b as (possible) result of the power of actor a that is the issue. With this form of power the actions of actor a are the main point and these actions of actor a may be considered as a result but at the same time may also be seen as an expression of his power.[12] This can be made clear in the following example:
During a discussion in the House of Representatives one of the representatives asks the cabinet minister a relevant question. Without offering any further explanation the cabinet minister refuses to answer this question.

In this example the cabinet minister exercises power over the representative by failing to do something. This does not directly influence the representative's actions. This second form of power can be presented more generally by means of the following model:

(II) An *actor a* is capable of or tries to do or refrain from doing something, whereas it is clear to him that an *actor b* demands or can reasonably demand that he should not do so, respectively should not refrain from doing so.

In explanation of this model it should first be stated that it is not essential - if power is to be the issue - for actor b to demand explicitly that actor a should or should not do something. It is sufficient for the demand made to be reasonable. Let us illustrate this by using the example above. It might be that the representative is too timid to bring to the cabinet minister's notice the fact that the latter has disregarded his question. We are inclined to state that this does not alter the cabinet minister's use of power. It is altogether reasonable that the representative should demand an answer to his question.

On the other hand we can not only speak of power when the demand actor b makes of actor a is reasonable. If the demand made by actor b is unreasonable he in fact exercises power over actor a. A refusal by actor a to concede to this demand may best be characterised as a clash of power against power.

Both forms of power - differentiated above - are brought out in our description of power:

(5) Power is the capability or the means to achieve one's objectives against the will of another person.

By speaking in general of 'the achieving of one's objectives'[13] both forms of power are done justice. For one's goals may be served by influencing another person's actions, but also by merely doing or refraining from doing something oneself (against another person's will).

Having now curtailed and enlarged the concept of power it is now time to define it. The demands of technical measurement, rather than theoretical concerns have been the cause of our examining only one aspect of power.

The word 'power' is often considered to have two meanings, viz. that of 'a capability' and that of 'a means'.[14] In colloquial language we use power in the sense of a *capability* when we say that someone *has* power. By this we mean that, whenever he wants to do so, he will succeed in achieving his goals against the will of others.[15] In this context one could speak of the *power of control.*[16] However, success is not guaranteed when power is seen as a means. For there are both effective and non-effective means. In colloquial language we consider power as a *means* when we say that someone *exercises* power. By this we mean that a person tries to achieve his objectives against the will of others. In this case we might speak of the *exercise of power*.

The question as to whether someone explicitly exercises power in a political discussion, can, in our opinion, be answered by analyzing his statements. However, in this way it is far more difficult if not impossible to determine to what extent someone possesses power, or in other words, to what extent a political discussion is decided on the basis of power. For a person may in fact act in anticipation of another person's power,

although this cannot be derived from his statements. Suppose a cabinet minister and a representative disagree. When, after a discussion, the cabinet minister is put in the right without having explicitly exercised power to achieve this, then the representative may have been convinced by the arguments given by the cabinet minister. However, the cabinet minister's position of power may play a part as well. It is just possible that the representative only intends to serve his own political ambitions and agrees with the minister knowing that the cabinet minister's vote is important when it comes to the choice of a successor to replace an under-secretary who is about to depart.
For the reason mentioned above we will only pay attention to a curtailed meaning of the concept of power by considering it exclusively as a means and not as a capability. In other words we will only offer a measuring-instrument for power in the sense of the *exercise of power*. The description runs as follows:

(6) Power in the sense of the exercise of power is, to the actor, a means of achieving his objectives against the will of other persons.

In the previous section (sound) argumentation was related to the concept of communicative acting first put forward by Habermas. Likewise, the exercise of power may be related to the type of acting that in Habermas' opinion is complementary to this, viz. *strategic acting*. Put in the terminology worked out above, a person acts strategically when he does not try to achieve his goals by using the soundest possible arguments, but by other means like the use of money (and even bribery), coercion and deception. In other words, when someone acts strategically he exercises power. In order to prevent misunderstandings, it is important to bear in mind from the start that not every violation of these discussion rules can be interpreted as a sign of the exercise of power. Section 7 will go into this further.

Finally, it is important to note that political rationality is related to both communicative and strategic rationality. When we, for example, define political rationality as a means-end rationality, then political ends can vary e.g. from solving problems in society to enlarging ones own power. In both cases communicative rational acting is necessary when the end can only be reached on the basis of valid knowledge and/or the voluntary cooperation of other people. And in the case that ends can only be striven for in competition with opponents, it can be wise to act strategically.

6.4 Defining and specifying the question

In the introduction we expressed our aim of developing a measuring-instrument in order to be able to answer the question to what extent a political discussion is a matter of sound argumentation or a matter of power. The determination of the terms sound argumentation and power in the previous two sections enables us now to define and specifiy this question in more detail.

The question of to what extent sound argumentation is found in a political discussion will be understood as the question to what extent the line of reasoning in a political discussion develops according to discussion rules (to be determined in detail). In other words in this question the issue is the procedural rationality of a political discussion.

We will only look at power as a means to achieve objectives against the will of other actors (we are speaking of the exercise of power then). The question of to what extent power plays a part in a political discussion will therefore be considered as the question of to what extent power is being exercised.[17] In other words we can also speak of the question of to what extent the participants in a discussion act strategically. A person acts strategically when he does not try to achieve his objectives by the soundest possible arguments but by means like the use of money (or even bribery), coercion and deception.

The measuring-instrument to be developed next will therefore have to give a definite answer to the following question:

(1) To what extent is a political discussion either of a strategic or of a procedural, rational nature?

In this question there is a certain incongruity. In the case of strategic acting a person *tries* to impose his will on others, whereas we find procedural rationality according to the extent to which a person *succeeds* in framing his statements regarding form and style according to discussion rules (yet to be determined). If such a difference is not desirable it is possible to ignore the procedural rationality, and instead look at the extent to which a person acted communicatively. A person acts communicatively when he tries to come to an agreement with others on the basis of the most plausible arguments. Whether he succeeds remains undecided, just as with strategic acting. In that case an alternative question could read as follows:

(2) To what extent is a political discussion of a strategic or of a communicative nature?

We will find that the measuring-instrument developed hereafter clears the way for the answer to this second question.

6.5 Rules for discussion

If the quality of an argumentation is to be assessed by the concept of procedural rationality, this concept has to be operationalised in more detail. This will be done in two moves. In this chapter we shall first formulate a number of rules for discussions - including political discussions - with which a rational argumentation procedure will have to comply. Subsequently, in the following section we shall point out how to operationalise these rules further on the basis of possible violations of these rules.

In section 2 it was stated that the procedural rationality of an argumentation can be determined by means of an assessment of the extent to which the manner of reasoning has been framed according to discussion rules (still to be determined). The starting-point for these rules is that they have to fit within the basic concept of communicative rationality chosen by us. Then the following three *principles* can be drawn up:[18]

1. one must speak and act on the basis of arguments;
2. the arguments one presents must be as sound as possible with respect to form and style;
3. acting and speaking must be aimed at coming to an agreement with others.

The discussion rules to be formulated next will have to comply with one or more of these principles.

In the literature of divergent disciplines such as political science, jurisprudence, philology and scientific philosophy, one finds suggested rules for the standardization of discussions (Alexy, 1981; Van Eemeren and Grootendorst, 1983; Grice, 1981; Habermas, 1973; Naess, 1978; Posner, 1974; Vedung, 1982). It would go beyond scope of this chapter to give a detailed account of this. We confine ourselves to formulating a number of rules drawn up on the basis of this literature. The entire of body of these rules forms -so to speak- a model procedure or in other words an ideal model, on the basis of which an actual discussion can be judged.[19] Here it may be stated that the more have been complied with these rules, the greater the procedural rationality will be.

First a survey of these rules will be given and some of the rules will be explained. Next a short explanation will be given on how the discussion rules relate to the principles formulated earlier.

A model procedure for discussions

(1) A committed attitude

(1.1) One is committed to the objective of the discussion
(1.2) One is committed to the things one has said and implied therewith
(1.3) One is committed to the arguments being solid

(2) Accountability

(2.1) Every participant in a discussion is to support his statements with the help of arguments, when other participants (may be expected to) demand this, unless he gives plausible reasons justifying a refusal.
(2.2) When one doubts the arguments relating to the point of view of another participant in the discussion, one may only challenge these if one gives counter-arguments.

(3) Consistency

The participants in a discussion act and speak in a consistent way.
(3.1) The participants in a discussion are not allowed to contradict themselves.
(3.2) The participants in a discussion are consequent.

(4) Relevancy

(4.1) The arguments one gives and the information going with them, must be relevant.
(4.2) When making a statement that (apparently) does not refer to the statements and arguments which are the subject of the discussion, one has to state one's reasons for making this statement, if other participants (may be likely to) expect this.

(5) Objectivity

The participants in a discussion are to adopt an objective attitude.
(5.1) One is not allowed prevaricate.
(5.2) One is not allowed to ascribe another persons points of view that they do not support.
(5.3) The points of view held must not be tendentious due to ambiguity.
(5.4) The participants in a discussion are not allowed to present their own contribution(s) to the discussion tendentiously, by means of incorrect or incomplete information.
(5.5) One should not become personal.

(6) Openness
The participants in a discussion must see to it that the discussion is open to others and to discussion contributions from others.
(6.1) It must be possible for everyone (to the same extent) to take part in the discussion.
(6.2) The participants in a discussion are allowed to raise any point of view and advance any information they consider relevant for the defence or challenge of a certain point of view.
(6.3) One is allowed to challenge each statement brought in by another participant to the discussion to justify or refute the expression of an opinion.
(6.4) The participants in a discussion are to provide as much information as necessary (for the aim of the discussion at that moment).

A number of these rules will now be explained briefly.[20]When taking part in a discussion, one must commit oneself to directing one's contribution to the objective of the discussion (rule 1.1). In the case of an argumentative discussion this implies -among other things- that one should at least try to make a point of view acceptable or unacceptable. To give an example of withdrawal, one withdraws from the objective of a discussion when only speaking to spin out time, i.e in order to hold up a decision being taken. One must be prepared to let one's actions or convictions depend on the solidity of the pro- and counter arguments (rule 1.3). This means that one is obliged to retract a point of view if it has been countered effectively and that one should stop doubting a point of view once it has been defended effectively (cf. Van Eemeren and Grootendorst, 1983, p. 174).[21]

A communicative-rational participant in a discussion can present arguments when asked for (rule 2.1). And he will give them without being asked in cases where it seems likely that the other participants will demand supporting evidence for his point of view. This is the case when one challenges a generally accepted statement. Alexy indicates that a discussion will lead nowhere if one of the participants calls in question every argument supporting a point of view and keeps asking endlessly for a further motivation (Alexy, 1981, p. 244). Rule 2.2 aims at preventing such practice by prescribing that one is only allowed to challenge the arguments of another discussion-participant, when mentioning counter arguments oneself.

The rule of consistency (3) means that a discussion contribution should be free from inner contradiction. Rule 3.1 refers to the logic of a discussion contribution. The statements one makes should not contradict one another. Moreover, in rule 3.2 it states that one is to be consequent. In accordance with Alexy, we can state that a person is consistent when he abides by the following code: 'when giving the object A a predicate F, one must be prepared to apply F with all other objects that correspond to A in all relevant respects' (Alexy, 1978, p. 235).

If an argument is maximally relevant (rule 4), this means that the plausibility of the argument is transmitted in its entirety to the conclusion. An irrelevant argument does not contribute in any way to the plausibility of the conclusion, however acceptable the statement serving as the argument may be (see Govier, 1985, p. 102 and Schellens, 1985, p. 74). An illustration of this would be the statement, that people have availed themselves on a large scale of an investment subsidy, is not a relevant argument for the conclusion that the investment subsidy has resulted in extra investments. For the fact that an investment subsidy is used, does not imply that this subsidy has caused investments that would not otherwise have been done without this subsidy. As pointed out by Alexy, it is not always immediately clear to all discussion-participants whether the arguments and information given are relevant. His view is that judging relevancy should be left to the participants involved in the discussion (Alexy, 1981, p. 244-5). For this reason rule 4.2 is included. One must motivate why one makes (apparently) non-relevant statements if other discussion-participants may expect you to.

An objective attitude in a discussion (rule 5) means in the first place that one devotes oneself 'for the sake of the cause'. This implies in fact that one does one's level best to achieve an as true or as correct an outcome to the discussion as possible. Tendentious and incorrect statements have no place here. An objective attitude means, in the second place, that one should not become personal in order to influence the outcome of a discussion. This means that one must not treat the other discussion-participants personally and that one must not 'throw one's own person onto the battle field'. Not being allowed to treat other discussion-participants personally, has mixed implications. One is not allowed to attack or threaten other participants, nor is one allowed to give them the advantage. When 'throwing one's own person onto the battle field', one refrains from giving arguments. Instead, one parades one's own qualities or one guarantees personally the correctness of a point of view.

The rule of openness (6) aims, in the first instance, at ensuring that everyone has an equal chance to take part in the discussion and is treated in the same way. This is allied to a second aim, viz. that levelling criticism may occur without obstruction and is to be stimulated. In compliance with rule 6, an important condition is that the contributions to the discussion are not made under coercion and are not restricted. One should not refrain from publishing a point of view or some information for fear of reprisals.

In general, the discussion rules adhere to the principles, formulated earlier, in the following three ways:

1. Speaking and acting on the basis of arguments is prescribed thereby stimulating coming to an agreement with each other (rule 1.3, rule 2 and rule 4.2).
2. The quality of form and style of the arguments is prescribed thus stimulating people to come to an agreement with one other (rule 3, rule 4.1, rule 5 and rule 6).
3. The objective of coming to an agreement with one another is prescribed (rule 1.1 and rule 1.2).

6.6 A method of operationalising the rules for discussions

A method of operationalising the rules for (political) discussions mentioned in the previous section, is to outline concrete situations for each rule showing how the rule is being violated.

For this purpose, extensive literature on fallacies can be made use of (see e.g. Damer, 1980; Van Eemeren and Grootendorst, 1986; Engel, 1985; Fearnside/Holther, 1963; Gilbert, 1986; Hamblin, 1980; Larrabee, 1952; and Pirie, 1985).[22] As indicated by Van Eemeren and Grootendorst, these fallacies can, in principle, be considered as violations of discussion rules (1983, p. 177). Furthermore these fallacies can also play a part when critically judging a system of rules for discussions, as drawn up in the previous section. The greater the number of fallacies that can be classified as a violation of one of its rules, the more satisfaction this system will give (cf Van Eemeren and Grootendorst, 1983, p. 151).

It is indeed possible to interpret the fallacies in the above mentioned literature as violations of one of the rules for discussion. To do this would be beyond the framework of this chapter.[23] Solely for the sake of illustration, a few selected fallacies will be examined, coupling one with each of the six rules for discussion. These six fallacies may be summarized as follows.

Denying a Common Starting Point
One casts doubt on a proposal which is one of the common starting-points. If one finds that, in a discussion, one is going to loose if one sticks to the initial starting-points, one may decide to deny these starting-points. Imagine a discussion in which one has to choose between two alternative ways of reaching a certain common policy objective. In this discussion a participant may decide to doubt the correctness of the policy objective if the way he has suggested is not considered to be the most effective. This, and similar ways of acting, are in defiance of the rule that one is committed to what one has said (rule 1.2).

'Slippery slope'
The slippery slope: one suggests unjustly, that holding a certain point of view will make matters go from bad to worse. One demonstrates that adopting a certain point of view results in adverse consequences by way of a string of related results, although this argument cannot be supported. This is the case in the following argumentation: "The proposal to increase the welfare benefits by 1,5% should not be approved. An increase in benefits may result in a situation whereby people on an allowance will be more reluctant to accept a job. Before you know where you are, nobody in this country will be working anymore. That is not what you want, that is not what I want. So let us refrain from increasing welfare benefits".

Confirming the Consequence
Confirming the consequence is a form of invalid reasoning in which a sufficient condition is regarded as a necessary condition. The following is an example of such reasoning: "Unemployment leads to criminality. Last year criminality rose considerably. Therefore unemployment has risen as well." The reasoning in the example does not hold good because criminality may have risen due to other causes. This and similar reasonings are in contradiction with the prevailing logic and are therefore a violation of the rule of consistency (rule 3.1).

Rejecting Alternatives
One supports a proposal by labelling alternatives as inferior. This method forms a violation of the rule of relevancy (rule 4.1).[24] Taken by itself summing up objections against other possibilities does not say anything about the quality of the proposal.

Partly Hidden Qualification
A limited claim is explicitly stated, but the intonation and/or structure of the sentence are such that a general claim is suggested. An example of this is to be found in the sentence structure "for practically every single case p holds good" a limited claim is made ("practically"), whereas the part "for every single case" suggests a general validity. Applying partly hidden qualifications results in points of view becoming tendentious due to ambiguity (rule 5.3).

Apriorism
One starts from certain starting-points or principles. Facts introduced as counter arguments against them are rejected, because they are not in conformity with these starting-points or principles. When acting like this one is in fact not open in a discussion contribution from others (rule 6).

As stated, these and other fallacies can be applied in the operationalisation of discussion rules. Of course this does not alter the fact that, in practice, these rules may be violated in more ways than just by fallacies mentioned in the literature. These fallacies only offer a clue for the operationalisation of the rules for discussions.

6.7 The assessment of discussions

On the basis of the rules for discussions, and their operationalisation (with the help of possible violations of these rules), political (and also other) discussions may be assessed to see to what extent to which either (sound) argumentation plays a part or power is being exercised.

The *quality of the argumentation* in a discussion is judged on the basis of its procedural rationality. The latter is greater when the rules for discussions are violated less often. The procedural rationality of a discussion - and of a separate discussion-contribution as well - may be quantified as follows:

$$PR = \frac{U - Uo}{U} \quad (1)$$

PR = the procedural rationality of a discussion (contribution)

U = the number of statements (arguments and conclusions)[25]
Uo = the number of different statements showing at least one violation of the rules for discussions

The *exercise of power* is considered to be strategic acting. When people are acting strategically in a discussion, this shows in the violation of the rules for political discussions. Both forms of power as differentiated in section 3 may be of influence here. One may get another person to do or refrain from doing something by providing him with false information (violation of rule 5.4). It is also possible to disregard something another person claims by not withdrawing a criticism of an expression of opinion although this has been effectively defended (violation of rule 1.3). A complication is, however, that not all violations of the rules for discussions can be seen simply as the result of strategic acting. This is only the case when it appears that one of the participants to the discussion does not intend to use the soundest possible arguments or when he applies pressure by means other than argumentation, such as the use of money (or even bribery), coercion and deception. In a discussion one is not always fully aware of the nature of the acts one commits. Often one is completely absorbed in the acting itself, not (immediately) reacting to the acts one commits. Therefore it is, in principle, possible for a person to act strategically without noticing it. In the heat of a discussion a person might demand 'point-blank' that someone else should put him in the right. This person may be so obsessed about being right, that he does not stop to think about the fact that demanding this is in fact a form of exercising power over the other person.

A violation of the discussion rules may result from communicative non-rational acting as well. When acting *communicative non-rationally*, one intends, just as in case of communicative rational acting, to come to agreement with other actors on the basis of as sound as possible arguments, but in the case of communicative non-rational acting one does not succeed very well. It may be, that one makes a mistake in logic in a complex pattern of reasoning without noticing it. All the same, it is possible to consider the violation of certain rules and the occurrence of certain fallacies, in general, as an expression of strategic acting. Generally speaking, the violations of rule 1 (a committed attitude) are active attempts to shirk an obligation one has entered into. And whether one adopts an objective or a non-objective attitude (rule 5) is a conscious choice. Therefore a non-objective discussion contribution may be seen as an expression of strategic acting. Similarly, it holds good that the violations of rule 6 (openness) are usually active attempts to keep other people or their contributions out of the discussion. A large number of violations of the other rules may be considered, quite simply, to be the result of strategic acting. Supporting this assertion extensively is not feasible within the frame of this chapter (see Pröpper, 1989). For the sake of illustration we refer here only to the 'slippery slope' fallacy dealt with in the previous section. When one unjustly implies that a point of view results in matters going from bad to worse, without supporting evidence, one is in fact acting tendentiously. One is acting strategically at that moment.

Summarizing, it can be stated that the extent to which a political (or any other) discussion is of a strategic nature, may be determined by establishing to what extent the rules for discussions are violated and by subsequently indicating whether or not the violations result from strategic acting. The extent to which a discussion is of a strategic nature, may be quantified as follows:

$$SCD = \frac{Uos}{U} \qquad (2)$$

SCD = the strategic component of a discussion (contribution)
U = the number of statements made (arguments + conclusions)
Uos = the number of different statements showing at least one violation of the rules for discussions as a result of strategic acting

So far we have shown how to provide an answer to the question of to what extent a political discussion is of a procedural rational or of a strategic nature. This was the first question formulated in section 4. The second question read: "To what extent is a political discussion either of a strategic or of a communicative nature?". How to determine to what extent a political discussion is of a strategic nature has already been determined, what remains is the determination of the extent to which a political discussion is of a communicative nature. A person acts communicatively when he tries to come to an agreement with other people on the basis of the soundest possible arguments. Whether he succeeds, is not important here. Communicative non-rational acting is a form of communicative acting too.

Communicative acting can be considered as the complementary concept of strategic acting. In other words, one is acting communicatively to the degree in which one is not acting strategically.[26] Therefore the extent to which a political discussion is of a communicative nature, may be quantified as follows:[27]

$$CCD = \frac{U - Uos}{U} \qquad (3)$$

CCD = the communicative component of a discussion (contribution)
U = the number of statements made (arguments and conclusions)
Uos = the number of different statements showing at least one violation of the rules for discussions as a result of strategic acting

6.8 An example of an application: empirical research into argumentation and the exercise of power

In the period 1986-1989 we investigated all the evaluation research into the application of The Investment Subsidies Act, called in Dutch the "Wet Investeringsrekening" (WIR), and the utilization of this research in the policy-making process.[28]

The aim of the WIR was to induce investments and it was made available to private enterprises from 1978 to 1988.

The *central issue* in our research is the following: 'To what extent does the quality of the argumentation in evaluation research offer an explanation for the extent to which evaluation research is utilized in the policy-making process at central government level, and for the quality of the argumentation in the utilization of evaluation research in this process?'

Our basic assumption is that policy evaluation can be seen as an argumentation.[29]We have studied a total of ten cases (one evaluation study with its utilization is considered as one research case).[30]

In this section we will summarize and briefly explain the central conclusions of our empirical research.

Conclusion 1: The quality of the argumentation in the evaluation research into the WIR is low.

This central conclusion is based on two other conclusions:

1.1 The level of the procedural rationality of the argumentation in the studied evaluation research into the WIR is low.
1.2 The argumentation in the studied evaluation research into the WIR is mainly a way of exercising power.

The first conclusion (1.1) is based essentially on the following observations. We find that less than half the statements in eight out of the ten investigated evaluation studies can be considered as being procedurally rational. An average of 65 % of the statements contain one or more violations of the procedural rules for discussion.

Our second conclusion (1.2) is mainly based on two observations. An average of 59 % of the statements in the research reports investigated contain one or more violations of the procedural rules for discussion, by employing strategic acting. Only two of the ten evaluation reports investigated can be considered mainly communicative, although even in these two reports power is exercised in a major part of the argumentation.

Conclusion 2: The quality of the argumentation in the utilization of the WIR evaluation research in the policy-making process is low.

We started from the following operational description of the utilization of evaluation research in the policy-making process: 'An individual or (administrative) organisation utilizes evaluation research in cases where one or more elements of the research are raised explicitly in the point of view of this individual or organisation or in the considerations preceding this point of view'. This means that adopting one or more elements of research is not the only way to utilize it. One can also be said to utilize research if one criticizes and rejects one or more elements of the research. Then the research is still of importance when testing a point of view or proposal. Conclusion 2 is based on two other conclusions:

2.1 The level of the procedural rationality of the argumentation in the utilization of the WIR evaluation research in the policy-making process is low.

2.2 The argumentation in the utilization of the WIR evaluation research in the policy-making process is to be considered mainly as a way of exercising power.

The first conclusion (2.1) is based essentially on the following observations. In the utilization of nine out of the ten investigated evaluation reports we have found that in half or less than half of all statements no violations of the procedural rules for discussion occurred in the argumentation. An average of 70 % of all statements contain one ore more violations of the procedural rules for discussion. Our conclusion, that the utilization of research we have investigated, is mainly a means of exercising power (2.2) is based on the following arguments. In nine out of ten cases the strategic component constitutes more than half of the argumentation. An average of 62 % of the statements can be classified as merely a way to exercise power.

Schedule 1 shows a survey of the extent to which the discussion rules were violated in the ten evaluation research projects investigated and in their utilization in the policy-making process.

Schedule 1: Frequency of the violation of the main rules for the model procedure for argumentation in ten evaluation research projects into The Investment Subsidies Act and in their utilization in the policy-making process, in absolute numbers and in percentages.

Main rules	Violations in the evaluation research		Violations in the utilization of the evaluation research	
rule 1 (a committed attitude)	0	0%	0	0%
rule 2 (accountability)	36	20%	53	19%
rule 3 (consistency)	10	6%	17	6%
rule 4 (relevancy)	18	10%	18	7%
rule 5 (objectivity)	112	63%	148	53%
rule 6 (openness)	1	1%	41	15%
TOTAL	177	100%	277	100%

The above shows clearly that, in particular, the rule of objectivity is most frequently violated. In the evaluation research studied this is shown by e.g. providing false, tendentious and misleading information, reporting selectively from other research projects, one-sided summing up of pros and cons and giving a wrong picture of the representativity of the research. Non-objectivity in the utilization of evaluation research in the policy-making process is shown by incorrect, incomplete or selective reproduction of the research outcomes and in abusing the researcher's authority.

In our opinion there is a simple explanation for the fact that in the evaluation research the rule of openness is only violated incidentally (1% of the violations), whereas it is found rather often in the utilization of the evaluation research (15% of the violations). In many cases the evaluation research is the 'opening move' in a discussion. Generally speaking, one does not have to react to other individuals' discussion contributions here. In the case of the utilization of the evaluation research the rule of openness is violated due to a disregard of certain (crucial) elements of the evaluation research and due to ignorance of each other's contributions in the phase of the political decision-making.

Conclusion 3: The extent to which the investigated evaluation research into the WIR is used in the policy-making process is considerable.[31]

We have measured the extent to which an evaluation study is used in the policy-making process on the basis of four main points: The number of different elements of the study which are used (conceptual model, statistic data, conclusion, recommendations); the number of conclusions and recommendations used; the number of categories of relevant actors who use the study; the number of people who used the study. On a zero to one scale, the extent to which the research is used has an average score of 0.76. In nine out of ten cases the score for the extent to which is is used is larger than 0.62 and increases to a maximum score of 1.

Conclusion 4: The quality of the argumentation of the investigated evaluation research into the WIR has no correlation with the extent to which this research is used in the policy-making process.

This central conclusion is based on two other conclusions:

4.1 The procedural rationality of the argumentation of the investigated evaluation research into the WIR has no correlation with the extent to which this research is used in the policy-making process ($R^2 = 0.001$, $t = 0.076$, at two-tailed test $P > 0.20$, $N = 10$).

4.2 The extent to which the investigated evaluation research can be considered as a means of exercising power has no correlation with the extent to which this research is used in the policy-making process ($R^2 = 0.003$, $t = 0.1605$, at two-tailed test $P > 0.20$, $N = 10$).

This result corresponds with previous research. E.g. research by Van de Vall and Patton shows an absence of correlation between quality features of research and the extent of utilization of this research in the policy-making process (Patton, 1978; Van de Vall, 1980).

Conclusion 5: The quality of the argumentation in the investigated evaluation research into the WIR has a strong correlation with the quality of the argumentation in the utilization of this research in the policy-making process.

This central conclusion is based on some other conclusions:

5.1 The procedural rationality of the argumentation of the investigated evaluation research into the WIR has a strong positive correlation with the procedural rationality of the argumentation in the utilization of this research in the policy-making process ($R^2 = 0.610$, $t = 3.534$, at two-tailed test $P < 0.01$, $N = 10$).

5.2 The procedural rationality of the argumentation of the investigated evaluation research into the WIR has a strong negative correlation with the extent to which the argumentation in the utilization of this research in the policy-making process can be considered as a means of exercising power ($R^2 = 0.499$, $t = -2.823$, at two-tailed test $P < 0.02$, $N=10$).[32]

5.3 The extent to which the argumentation of the investigated evaluation research can be considered as a means of exercising power has a strong positive correlation with the extent to which the argumentation in the utilization of this research in the policy-making process can be considered as a means of exercising power ($R^2 = 0.604$, $t = 3.494$, at two-tailed test $P < 0.01$, $N = 10$).

Conclusions 5.1, 5.2 en 5.3 agree with our hypotheses.[33] To support them, we have used the following three considerations:

1. A positive consideration: Evaluation research can serve as an example and can fulfil an educational function. When the argumentation in an evaluation study is procedurally rational, i.e. committed, motivated, consistent, relevant, objective and open, this can be adopted by the users of the evaluation research.
2. A negative consideration: A correct presentation is particularly important to the procedural rationality of an evaluation study. This means that a clear picture is given of a study's reliability and validity, and of the importance which should be attached to the conclusions of this study. If this is done properly, there is less opportunity for potential users of the study to misuse the research results.
3. If a user of an evaluation study feels that this study is mainly set up as a means of exercising power, this may be reason for him to use some means of exercising power himself.

On the basis of these theoretical considerations we might conclude, after all, that the quality of the argumentation of the investigated evaluation research into the WIR influences, the quality of the argumentation in the utilization of this research in the policy-making process.

Based on the research results, the evaluation research might be expected to contribute rationally to the policy-making and bring about the soundest possible argumentation rather than the exercise of power in political discussions. If this is the case then, the evaluation research itself must not fail to meet the standards of rationality. Yet, this holds good to a lesser extent for the evaluation research into the WIR at issue here.

6.9 Final remarks

This chapter gives an account of a research project investigating the exercise of power and the quality of the argumentation in evaluation research and in the utilization of this evaluation research in the policy-making process.

There is no reason to believe that the investigated evaluation research into the application of the Investment Subsidies Act in the Netherlands is representative for all evaluation research. Our study is mainly of an exploratory nature, and the results of this study can only help to formulate a few expectations with respect to evaluation research in general.

A great part of this chapter deals with a proposal for a measuring instrument in order to establish to what extent either power or sound argumentation is to be found in political discussions. It is important to bear in mind the scope and restrictions involved. It will not be sufficient to merely attempt an assessment of an argumentation according to its contents. However, it is possible to use it in answering the question of to what extent one can speak of sound argumentation in procedural terms. It does not have the scope to determine the extent to which a political discussion is decided by power. However, it may provide a positive solution to the question of to what extent power is exercised in political discussions.

Notes

1. The author wishes to thank prof.dr. A. Hoogerwerf and prof.dr. F.H. Van Eemeren for their helpful comments on an earlier draft.

2. See Machiavelli (1976) and Tijmes (1977). The latter refers to M. Weber, *Gesammelte Politische Schriften*, Tübingen, 1971, p. 506 and 507 (see p. 117).

3. Compare also the following statement by Guépin: "He who advises consultation, advises distance versus his own being right and at the same time a certain amount of scepticism. All too soon your ideal will be the sight of a group of ironical and smart gentlemen, who feel self-confident enough to dispute and pretend their being right" (Guépin, 1983, p. 19).

4. Hoogerwerf (1979) refers to Bertrand de Jouvenel, *De la Souveraineté*, A la Recherche du Bien Politique, Paris, 1955.

5. On this point there is no difference between empirical and normative science. Compare Fischer, 1982, p. 319.

6. If the 'received' social rules are taken as a starting-point, one follows the rules as actually accepted by a certain group in a concrete situation. However, it is also possible to 'surpass' these, by assuming a prescriptive system of 'desirable' rules. The latter is opted for in this article, although it is true tot say for a number of discussion rules (to be drawn up later) that they are generally or frequently accepted.

7. As indicated before, these discussion rules will be drawn up in a following section.

8. According to Ellemers this definition comes very near to Weber's well-known definition: "Macht bedeutet jede Chance, innerhalb einer sozialen Beziehung den

eigenen Willen auch gegen Widerstreben durchzusetzen, gleichviel worauf diese Chance beruht" (Ellemers, 1978, p.25).
See also Lasswell and Kaplan: 'Power is a relationship in which one person or group is able to determine the actions of another in the directions of the former's own ends' (1950, p. 74) and see also the description of Dahl: 'A has power over B to the extent that he can get B to do something that B would not otherwise do' (1957).

9. Here we link up with Connolly. This author puts it as follows: "A does not exercise power over B when B is convinced by arguments given in good faith by A. B chooses freely the alternative directed by A" (Connolly, 1974, p. 88). See also Lukes, 1974, p. 32/33.

10. For the concept of authority has a wider effect. One may e.g. also speak of 'the authority over a ship'.

11. Compare Ellemers' position in this connection: "In my opinion authority is especially a *quality* of power or influence, viz. such power or influence that is accepted as *reasonable* and *right*" (Ellemers, 1976, p. 27). Also see Hoogerwerf, 1978, p. 90.

12. e.g. this is the case when we say that a person acts arbitrarily.

13. It should be noted here that these goals cannot only be achieved through the active participation of person a, but also by person b's anticipating the position of power held by of person a.

14. See also Lukes, 1974, p. 12.

15. Literature mentions power as 'potential influence' if one sees power as a capability. See e.g. Hoogerwerf, 1978, p. 89.

16. Also refer to the following statement by Hoogerwerf: "Effective exercise of power is often called control" (Hoogerwerf, 1978, p. 85).

17. If we compare the power measuring-instrument to be developed with a number of known measuring-methods in the literature, such as formal position method, the network method or informal position method, the participation method, the decision-making method and the reputation method (see e.g. Van Schendelen, 1976), there is a difference in what is to be measured. By means of these latter methods attempts can be made to measure power as a capability (potential influence) and/or effected power as a capability (factual influence). The argumentation-theoretical measuring-instrument in this paper tries to measure exercise of power.

18. These principles can be derived directly from the meaning of the concept of communicative rationality. The better one succeeds in coming to an agreement with

others on the basis of the most plausible arguments, the greater the communicative rationality will be.

19. A similar procedure is also followed by Van Eemeren and Grootendorst (1983).

20. The literature mentioned contains yet another set of rules. The nature of these rules will be explained briefly and the reasons for their not being dealt with.
Rules are mentioned covering the communicative competence of participants to a discussion and linguistic usage. These rules are difficult to operationalise and are more the subject of a linguistic than of a political study. To a certain extent this holds good as well for rules concerning the clearness of discussion contributions. These rules are also considered to include the rule that statements should not be ambiguous, it will be found that this approach of clearness is contained in rule 5.3. The rule that participants to a discussion should be veracious, is shown in rule 5.4 and for the rest can be ignored.

21. A point of view has been defended effectively when this has taken place by means of cogent arguments, that form part of the common principles. Cf Van Eemeren a.o., 1986, p. 91.

22. A fallacy may be considered to be a false or deceptive reason or reasoning.

23. See Pröpper, 1989, a survey of more than 70 fallacies that may be characterised as violations of various discussion rules.

24. In a few exceptional cases we cannot speak of a fallacy. In that case the following two conditions have to be fulfilled:
    1. the group of alternatives one scrutinizes is limited and known;
    2. it is indicated that the objections against the rejected alternatives do not apply to the proposal that is defended.

25. It seems as if by expliciting implicit statements, one can make more statements, resulting in a higher score. This is not the case if the following is observed when counting the statements:
    - two or more premises in a reasoning together forming one support for a conclusion, are counted as one argument and therefore as one statement;
    - arguments remaining implicit are counted as well.

26. This is a simplification. Essentially, strategic acts always have a communicative aspect attached to them. This can be illustrated by the following example. When person a acts strategically by making person b, under duress, do something against his will, the acting of person a is also communicative in the sense that he is aiming at arriving at an agreement with person b on the meaning of the latter's message. For person a only reaches his goal when person b understands his task and the consequences should he not carry out the assignment of person a.

27. A researcher may have strong doubts about whether a violation of the rules for political discussions results from communicative non-rational or from strategic acting. This means that a component of the discussion may remain non-concrete as a result of doubt (?CD). This component may be expressed as follows: ?CD = Uot / U, Uot meaning the number of different statements showing one or more violations of the rules for discussions, for which applies that the researcher doubts whether this results from communicative non-rational or from strategic acting. In that case the communicative component is as follows: CCD = {U - (Uos + Uot)} / U.

28. All the evaluation research carried out at the request of the Dutch government. See Pröpper, 1989.

29. See Ahonen ('Public Policy Evaluation as Discourse'), House ('The Logic of Evaluative Argument') en Meehan ('Reasoned Argument in Social Science, Linking Research to Policy'), and also Cronbach (1980), Dunn (1981, 1982, 1986), Fischer (1980, 1982, 1986), Garland (1985), Graham (1986), Milligan (1980), Mitroff and Mason (1982) and Yearley (1986). Milligan says the following: "Is it possible to seperate the evaluation or the evaluating from the reasoning? I do not think so. In deliberating we cannot first state our values and then reason about them: the two activities are inextricably mixed up. We evaluate in the process of forming reasons and in the process of relating different reasons to each other. Our evaluations are often shown by the way we reason." (Milligan, 1980, p. 93).

30. The summary or a summarizing chapter of the research report has been the basis for a reconstruction and assessment of the argumentation research. The argumentation in the utilization of the evaluation research in the policy-making process has been based on official and political public writings.

31. This conclusion is not very suprising. Our study focuses especially on the way in which evaluation research is used. In order to investigate this method of use, we have more or less incorporated a guarantee, that the research is used. We have only selected research that was done at the request of the Dutch government.

32. The greater the strategic part of an argumentation, the less its quality will be.

33. In this brief overview of our research we refrain from mentioning our hypotheses, because they match our conclusions.

References

Ahonen, P., (1985)
Public Policy Evaluation as Discourse, Helsinki, (2e)
Alexy, R., (1981)
Theorie der juristischen Argumentation, Die Theorie des rationalen Diskurses als Theorie der juristischen Begründung, Frankfurt am Main.
Banner, D.K., a.o., (1975)
The Politics of Social Program Evaluation, Cambridge.
Connoly, W.E., (1974).
The terms of political discourse, Lexington.
Cronbach, L.J., a.o., (1980)
Toward Reform of Program Evaluation, San Fransisco.
Dahl, R.A.,
The Concept of Power, in: *Behavioral Science*, 2, 1957, pp. 201-205
Damer, T.E., (1980)
Attacking Faulty Reasoning, Belmont.
Doorn, J.J.A., van, en C.J. Lammers, (1976) (in Dutch)
Modern Sociology, Utrecht/Antwerpen.
Dunn, W.N., (1981)
Public Policy Analysis, An Introduction, New York.
Dunn, W.N.,
Reforms as Arguments, in: *Knowledge: Creation, Diffusion, Utilization,* 3, 3, 1982, pp. 293-326
Dunn, W.N, (ed.), (1986)
Policy Analysis: Perspectives, Concepts and Methods, Greenwich/London.
Eemeren, F.H., van e.a., (1978) (in Dutch)
Argumentationtheory, Utrecht/Antwerpen.
Eemeren, F.H., van, e.a., (1986) (in Dutch)
Argumentationstudy 2: Fallacies, Groningen.
Eemeren, F.H., van, en R. Grootendorst, (1983)
Speech Acts in Argumentative Discussions, A Theoretical Model for the Analysis of Discussions Directed towards Solving Conflicts of Opinion, Dordrecht.
Ellemers, J.E., Power and Authority in: A. Hoogerwerf (red.),
Explorations in de Politics 1, Alphen a/d Rijn, 1976, pp. 25-36 (in Dutch)
Engel, M., *With Good Reason, An Introduction to Informal Fallacies,* New York.
Fearnside, W.W., and W.B. Holthern. (1963)
Drogreden of argument, Utrecht.
Gilbert, M.A., (1979)
How to Win an Argument, New York.
Govier, T., (1985)
A Practical Study of Argument, Belmont.

Fischer, F., (1985)
Critical Evaluation of Public Policy, A Methodological Case Study, Paper IPSA, Paris.
Fischer, F.,
Politics, Values and Public Policy, The Problem of Methodology, Boulder Colorado, 1980
Fischer, F.,
Policy Evaluation, Integrating Empirical and Normative Judgements, in: *Evaluation Studies Review Annual,* 7, 1982, p. 313-324
Fischer, F., (1986)
Practical Discourse in Policy Argumentation, in: W.N. Dunn (ed.), *Policy Analysis: Perspectives, Concepts, and Methods,* Greenwich/London, p. 315-332
Garland, D., (1985)
Politics and Policy in Criminological Discourse: A Study of Tendentious Reasoning and Rhetoric, in: *International Journal of the Sociology of Law,* 13, pp. 1-33
Graham, G.J., (1986)
Ethics, Rhetoric and the Evaluation of Public Policy Consequenses, in: W.N. Dunn (ed.),*Policy Analysis: Perspectives, Concepts, and Methods,* Greenwich/London, pp. 301-314
Grice, H.P.,
Logic and Conversation, in: P. Cole and J.L. Morgan (eds.), *Syntax and Semantics 3: Speech Acts,* New York, 1975, pp. 43-58
Guépin, J.P., 1983 (in Dutch)
The civilized society, Amsterdam.
Habermas, J., (1981)
Theorie des kommunikativen Handelns, Frankfurt am Main, (2 Bände)
Habermas, J., (1983)
Diskursethik - Notizen zu einen Begründigungsprogram, in: J. Habermas, *Moralbewusstein und kommunikatives Handelns,* Frankfurt am Main, pp. 53-127
Habermas, J., (1973)
Wahrheitstheorien, in: H. Fahrenbach, (Hrsg.), *Wirklichkeit und Reflexion,* Pfüllingen, p. 211-265
Hamblin. C.L., (1970)
Fallacies, London.
Hoogerwerf, A., (1979) (in Dutch)
Political Science, Concepts and Problems, Alphen aan de Rijn.
Hoogerwerf, A., (1986 (in Dutch)
Visions of the Political Elite, Amsterdam.
House, E.R., (1977)
The Logic of Evaluative Argument, Los Angeles.
Larrabee, H.A., (1952)
Benthams's Handbook of Political Fallacies, Baltimore.
Lasswell, H.D., and A. Kaplan, (1950)
Power and Society, New Haven.

Lukes, S., (1974)
Power: A Radical View, London.
Machiavelli, N., (1961) (translation by G.Bull)
The Prince, Middlesex.
Meehan, E.j., (1981)
Reasoned Argument in Social Science, Linking Research to Policy, Westport/London.
Milligan, D., (1980)
Reasoning and the Explanation of Action, Sussex.
Mitroff, I.I. and R.O. Mason, and V.P. Barbara, (1982)
Policy as Argument - A Logic for Ill Structured Decision-problems, in: *Management Science*, 28, 12, pp. 1391-1404
Naess, A., (1978) (in Dutch)
Argumentationtheory, Baarn.
Patton, M.Q., (1978)
Utilization-Focused Evaluation, Beverly Hills.
Pirie, M., (1985)
The Book of the Fallacy, A Training Manual for Intellectual Subversives, London.
Posner, R., (1974)
Diskurs als Mittel der Aufklärung, Zur Theorie der rationalen Kommunikation bei Habermas and Albert, in: *Linguïstik und Sprachphilosphy*, pp. 280-303
Pröpper, I.M.A.M., (1989) (in Dutch)
Argumentation and the Exercise of Power in Research and Policy-making, Enschede.
Rapoport, A., (1980)
Various Meanings of 'Rational Political Decisions', in L. Lewin, and E. Vedung (eds.), *Politics as Rational Political Action*, Dordrecht, pp. 39-59
Schellens, P.J., (1985) (in Dutch)
Sound Arguments, Dordrecht.
Schendelen, M., van, (1976) (in Dutch)
Central themes in Political Science, Meppel.
Tijmes, P., (1983) (in Dutch)
Politics can be Discussed Rationally, Meppel.
Vall, M., van de, and C.Bolas., (1980)
Applied Social Discipline Research or Social Policy Research, The Emergence of a Professional Paradigm in Sociological Research, in: *American Sociologist*, 15, 1980, pp. 128-137
Vedung, E., (1982)
Political reasoning, Beverly Hills.
Yearley, S., (1986)
Interactive Orientation and Argumentation in Scientific Texts, in: J. Law (ed.), *Power Action and Belief, A New Sociologiy of Knowledge*, London, pp. 132-157

PART IV

Ethical Problems with Developments in Science and Technology

7 SCIENTIFIC KNOWLEDGE, DISCOURSE ETHICS, AND CONSENSUS FORMATION ON PUBLIC POLICY ISSUES

Matthias Kettner
Johann Wolfgang Goethe University, Frankfurt am Main, Federal Republic of Germany

7.1 Introduction

Discourse ethics (communicative ethics) as developed by K.-O. Apel and J. Habermas postulates a number of procedural conditions for rational consensus formation with regard both to normative and factual issues. In the first part of this chapter I present an outline of the theoretical and the practical level of discourse ethics. I claim that discourse ethics provides a critical yardstick for evaluating processes of consensus formation about public policy decisions. I introduce the normative concept of the "consensual etiology" of an existing institutional arrangement. In the second part of the chapter I apply the concept of a consensual etiology in order to evaluate critically the history of public consensus formation about the civil use of nuclear power. Public consensus formation about nuclear power falls short of certain morally relevant constraints. Furthermore, I discuss how political decision making concerning nuclear power relies on "mandated science", i.e. science used or interpreted for the purposes of making policy, as a questionable surrogate of morally required consensus.

Discourse ethics is a normative moral theory developed by Karl-Otto Apel (Apel 1972, 1978, 1979, 1982, 1988, 1990) and Jürgen Habermas (1989, 1990). Discourse ethics has been characterized as deontological, procedural, universalistic, and cognitive (Habermas 1989). More important than these labels, which are only partially correct, is the fact that discourse ethics is a two-level theory. On the first "theoretical" level, discourse ethics is a specific programm of metaethics and normative philosophical ethics. its objective is to specify an justify prescriptive contents that can then serve as regulative ideas, i.e. as rationally justified operative idealizations, on the second level ("level of practical discourse"). It is up to philosophy to theoretically specify the concept of practical discourse; practical application of this concept in the non-ideal actual world proceeds through social practices of persons whose moral life is not up for philosophy to control. Discourse ethics thus distinguishes between the role of the moral philosopher as theorist and the role of someone involved in the resolution of actual moral problems. The main point of this distinction is to avoid intellectual paternalism. In the paragraphs 7.2 and 7.3 I will give a condensed description of the two levels of discourse ethics.

7.2 The theoretical level of discourse ethics: deriving prescriptive contents from presuppositions of argumentation

On the philosophical or theoretical level, discourse ethics attempts to delineate a range of prescriptive contents for which a rationally ultimate foundation can be expected, to spell out such contents, and to establish their presumably universal validity by ascertaining their rationally ultimate reasons. Such reasons, it is claimed by Apel, can be found in the very activity of argumentation itself, i.e. they can be ascertained when people involved in argumentation (e.g. about the truth of a proposition p) reflect on the operative presuppositions of their speech acts in argumentation, presuppositions without which the sense and point of their speech acts would suffer or even be lost. Such presuppositions are thus non-contingently related to argumentation.

Apel calls such non-contingent presuppositions of argumentation "transcendental-pragmatic" presuppositions: "pragmatic" in as much as they pertain to and are operative in a communicative *praxis*, namely argumentation; "transcendental" in as much as commitment to those presuppositions is unavoidable for any speaker intending his speech acts under those action-concepts that belong to our concept of communicative rationality, i.e. to a concept that is itself *indispensable* for and irreplaceable in every possible discursive world.[1]

In argumentation, we are dialogically and cooperatively trying to zero in on the *best reasons* (e.g. to hold true, that p., or to hold right, that S ought to do A). Hence we are trying to furnish reasons that we assume, under ideal conditions, would convince everybody competent with regard to he disputed issue of why such and such ought rationally be held to be valid,or invalid. Argumentation, then, is a dialogical procedure whose outcome tends to not so much reflect unequal powers, differences in social status of divergent intellectual abilities of the participants but rather the force of the better argument only. Argumentation that operates under this regulative ideal, Apel and Habermas call "discourse".

Apel claims that there are transcendental-pragmatic presuppositions of argumentation there are morally relevant and have universal normative validity. Consider, e.g., the fact that participants in discourse necessarily presuppose, with regard to all beings capable of speech and argumentation, their reciprocal recognition as free and equal persons, namely as persons who are equally entitled to, and equally free to, express consent or dissent to discursively raised validity claims entirely on the basis of their rational evaluations of reasons and arguments. Furthermore, in argumentation over validity claims we presuppose that intrinsic aim of our activity is the settling of disputed validity claims by rational agreement reached entirely through uncoerced dialogue. Although dialogue in the actual world is always bound to be dialogue within a limited community of particular persons, the audience that is being addressed in terms of counterfactual intentions is a "virtually infinite community of communication" (Apel 1981). Apel has made this point in terms that draw on Charles S.Peirce's seminal idea of an infinite community of investigators.

Both these morally relevant presuppositions are strongly counterfactual: Though any real episode of argumentation will conform to these presuppositions in virtue its being an episode of argumentation, yet neither will it always be evident to the participants that they make these presuppositions, nor can they be sure that these presuppositions are in fact sufficiently fulfilled within the confines of the participants' actual community of communication. However, if someone substantially doubts their fulfilment with regard to a particular episode of argumentation(say, a participant in a discourse concerning the truth of some proposition p learns that potentially valuable contributors have been shunned or that relevant evidence has been suppressed) then she has prima facie good reasons to withhold consent and to criticise the impaired communicative rationally embodied in that particular episode.

The nature of transcendental pragmatic presuppositions can be clarified by comparison to Kant's regulative ideas. The aim of this paper, however, is to apply rather than to theoretically defend discourse ethics. Hence I will not pursue this point here. Instead, I conclude my exposition of the first theoretical level of discourse ethics by briefly introducing some points that

Apel hopes (Apel 1987) can be defended by appeal to transcendental-pragmatic presuppositions of argumentation.

7.2.1 Discourse as product and as process

Ideally, the outcome of discourse will be a rational consensus. However, any particular consensus as the product of a concrete, historically situated and hence limited community of communication is fallible, and therefore must be in principle open for revision: if the communicative rationality that presumably was embodied in a specific consensus-oriented process can be shown to have been seriously flawed, or if new arguments pertinent to the contents of any particular consensus emerge, discourse has to be taken up again. Discourse is thus an open-ended project. The authority of any discursively reached consensus is the authority of rational agreement, i.e. any discursively reached consensus, despite its fallible nature, manifests that which has in its favour our best reasons for anyone: With regard to truth, any discursively reached consensus about the truth of some proposition p will manifest the best reasons for anyone to agree to p's being true (p is, then, what everyone ought rationally to believe); with regard to moral rightness, any discursively reached consensus about the moral rightness of some prescriptive contents n (such as, e.g., a particular norm of action) will manifest the best reasons for anyone to consider n's being morally binding on everyone (n is, then, what everyone ought morally to abide by).

Discourse ethics accounts for the social binding force of moral commitments only to the extent to which such commitments can be justified by appeal to universalisably good reasons. Moral commitments, of course, may and do rest on a host of other force-giving sources. And not every moral commitment, of course, is by its content addressed to everyone. The accounts of moral psychologists and sociologists abound with explanations of societal or intrapersonal sources and mechanisms which generate forces that the individual perceives as morally binding, and anthropologists have a lot to say about group-specific moral commitments. The contention of discourse ethics is, rather, to spell out the cognitive kernel of morality: if morality has a rational core at all, then that core will have to manifest itself in discursively reached consensus about the validity of prescriptive claims.

In the next section I introduce two principles that serve to pinpoint the Kantian heritage in discourse ethics.

7.2.2 Dialogical universalisability, and the practical discourse demand

Discourse ethics modifies the Kantian principle of universalisability into a principle of dialogical universalisability. Instead of asking what an individual moral agent could, or would, will, without self-contradiction, to be a universal maxim for all, constraint-free discourse among real people has to determine which norms or normative institutional arrangements can be freely accepted by everyone concerned.

The dialogical reading of Kant's categorical imperative yields the following criterion of moral rightness (normative validity): Every valid norm must satisfy the condition that all concerned persons or parties can accept the consequences and side-effects that its general observance can be anticipated to have for the satisfaction of everyone's interests(Habermas 1990). This idea captures and transforms the Kantian notion of a test for moral worthiness of maxims of action. Discourse ethics modifies the Kantian position in yet another important respect. In discourse ethics Kant's notion of the "good will" and the imperative side of Kant's categorical imperative are taken up in the notion of a morally based demand for a specific way of conflict resolution. Discourse ethics holds the following action-guiding principle as binding on every (real or possible) member of the (virtually infinite) community of communication: Whenever interests conflict, attempt to resort to practical discourse for resolving the issue!

On the first or theoretical level, discourse ethics postulates that the morally obligatory course of action is t try to transform conflicting interests into *competing claims* over such *norms* that would, if valid,regulate conflicts in a way that all people affected would be able to acknowledge as morally right. For the sake of brevity, I will refer to this demand as the "practical discourse demand". With this general action - guiding principle supplemented by the dialogical principle of universalisability, moral theory limits its domain of special authority. It is obvious that the practical discourse demand as it stands cannot be "applied" like a recipe to be followed or be "implemented" in domains of moral life by moral experts.

7.3 Prescriptive contents on the practical level of discourse ethics

7.3.1 The normative notion of a consensual etiology

On the second, or practical level, then, discourse ethics does not itself intend to generate concrete moral principles or norms of action or value-orientations. Instead, discourse ethics imposes a number of morally relevant constraints on the rational acceptability of any proposed or already entrenched concrete moral principles, norms of action, value-orientations, etc. On the strength of the practical discourse demand, their rational acceptability is constrained by their conforming or not conforming to practical discourse: Any prescriptive content commanding universal allegiance that could not have come into such a position via the route of dialogical universalisability will be disqualified for failing the practical discourse demand set up for moral rightness by discourse ethics.

I suggest we can best think of this demand in terms of a "consensual etiology": discourse ethics constrains the rational acceptability of any normative order, or institutional arrangement tied to a certain normative order, by requiring it either to have a consensual etiology, or to be sufficiently conceivable as an equivalent of something which could have, or might have had, a consensual etiology. The idea of a consensual etiology should not be misunderstood as a utopian idea. Rather, the idea of a consensual etiology of some established normative order or existing arrangement x is a critical yardstick for evaluating the moral acceptability of x. The moral credit x deserves depends crucially on an evaluative comparison between the potential of opportunities for practical discourse that were bound up with the emergence of x on the one hand, and how x really made its way into the actual world. to see how the practical discourse demand is relevant with regard to x in a certain episode of x's emergence is to see how the practical discourse demand could be attended to in that episode. How it is actually dealt with, or whether it is dealt with at all, has to be evaluated against this background. For instance, x's moral acceptability will be found to be impaired to the extent that absence of or distortions in the procedural conditions for practical discourse in relevant episodes can be explained as effected by agents with vested interests in the emergence of x.

To see the critical force of the practical discourse demand, let's take a closer look at the notion of practical discourse. In the following section, I will point out a number of constraints that provide content to the notion of practical discourse. The five constraints I wish to underline make clear important respects in which practical discourse differs from less demanding moral conversations and from practical deliberation in general. The constraints that I now address also have momentous political implications with regard to public policy making. From paragraph 7.4 on I will explore some of those implications specifically with regard to the issue of nuclear power.

7.3.2 The level of practical discourse: five morally relevant constraints

1. *The generality constraint*

First, practical discourse over an issue ought to be open to all competent speakers whose interests are or will be affected by regulations adopted to resolve the issue, I will call this feature of practical discourse the generality constraint. The generality constraint does not imply that any and every person affected will have to be heard; rather, what it implies is that the arguments put forward by the actual participants will have to be fairly representative of the arguments that all others concerned would or could present. The best way to ensure this is, of course, to make participation as encompassing as possible. Where the interests of future generations are at issue, and direct participation is impossible, proxy or surrogate participants will have to step into their argumentative place in the discourse. It is debatable whether the representational capacities that are institutionalized in the parliamentary systems of modern democracies are already sufficient for these tasks.

2. *Autonomous evaluation constraint*

Second, practical discourse provides its participants with symmetrical chances to introduce and challenge assertions, and to express their needs and wishes. This implies a principal non-paternalism. Practical discourse starts with the very terms in which the participants themselves construe the issue in question, their respective interests, and their moral commitments. Unlike objectifying moral theories such as, e.g. preference utilitarianism, practical discourse does not replace the concepts under which people really intend their being moral into some dogmatic format. This does not imply that each individual's needs and wishes expressed as interests go unchallenged. Concepts can be questioned, e.g. when they are descriptively inadequate. Individual as well as collectively shared values and ideals can be challenged, e.g. when they dogmatically rule out alternatives. Interests can be considered illegitimate, e.g. when the needs, wishes or beliefs on which they are based can be shown to be irrational. I will call this feature of free accessibility of critical evaluations under non-paternalistic conditions the autonomous evaluation constraint of practical discourse.

Related to the autonomous evaluation constraint are two further points, namely that practical discourse requires that the participants are subject neither to internal nor external coercion. I will address the absence of internal coercion first.

3. *Role taking constraint*
Participants must be able to adopt a hypothetical stance towards their own interests, values, needs, etc., as well as to those expressed by others. Internal coercion (e.g. strong neurotic fixations) will prevent people from adopting such a stance. To be capable of taking an interest in each others interest, and to be prepared to let one's own interests be radically questioned, calls for what G.H.Mead, L.Kohlberg (1990) and others have termed "ideal role taking". I will refer to it as the role taking constraint.

4. *Power neutrality constraint*
Absence of external coercion, on the other hand, means that existing power differentials between participants have to be bracketed or neutralized in some way so that they have no bearing within the cooperative pursuit of rational agreement through argumentation. For instance, a South-african farmer seriously debating the supposed moral rightness of slavery with his black servant cannot refer to his (still!) superior power position in order to prove his point, unless the discursive position is given up and left to deteriorate into a position of collective bargaining or, in the worst case, to strategic threatening. In the remainder of this chapter I will refer to this point as the power neutrality constraint.

5. *Transparency constraint*
The final feature of practical discourse that I wish to underline is the incompatibility of practical discourse with strategic action. Strategic action is success oriented action of an agent who treats others as limiting operation conditions of his or her actions, i.e. only as means to the agent's ends. Practical discourse requires participants to share a full understanding of their goals and intentions relevant to the issue. As strategic action, overt or covert, is incompatible with unreservedly cooperative pursuit of rational agreement, strategic action is outruled in practical discourse. Goals and intentions whose effectiveness require their not being shared (as, e.g., lying, betraying, deceiving, pretending, persuading, make believe) belong to latent strategic action, not to the consensual action that is required for practical discourse. I will call this incompatibility the transparency constraint of practical discourse. In the following paragraphs I will apply discourse ethics, as outlined so far, to the issue of nuclear power.

7.4 Scientific knowledge and consensus formation about nuclear power

National energy politics is an area of political decision making that obviously affects interests of the entire population. How momentous the decision, e.g., to embark on ambitious programs of nuclear power, in fact is should be apparent to virtually everyone, in the U.S. at least since the Harrisburg reactor incident of 1979, in Britain since the disclosure of massive hazards caused by leaking radioactive wastes in Sellafield, and in Europe generally since the Chernobyl catastrophe of 1986 that exposed many european countries to intense radioactive emissions. The political fall-out of Chernobyl has forced public policy makers in the energy sector to address resurgent anti-nuclear opposition and to confront a public increasingly aware of the potential costs and risks of an energy-path that for a long time had seemed for the majority to be unquestionably worth pursuing.

Looking at how political thematisation of momentous scientific and technological issues (such as nuclear power) unfolds historically, one detects a certain pattern: in an early phase, physical scientific and economic considerations are the dominant rationality aspect under which such issues are thematized; next, such considerations are in time "replaced in importance by sociopolitical questions, which, in turn eventually give way to moral and ethical issues that become crucial" (Del Sesto 1980, p.69).

We are presuably in this third stage now. This puts on the agenda a critical review of the fact that only now do moral issues surrounding nuclear power technology gain importance: in the beginnings of the natural history of nuclear power technology, were there no moral issues to be addressed? Or have they been with us all along, their importance only recently realized, and that literally by accident? And if so, is this a morally irrelevant fact about the contingencies of technological experience, public awareness and policy making? Or does this fact rather reveal morally relevant deficiencies of the kind of rationality that is embodied in technological experience, in public awareness and in policy making?

One way to raise these questions is to judge the natural history of nuclear power and its political thematisation against the normative model of practical discourse. In virtue of the *practical discourse demand* one ought to reconstrue the natural history of nuclear power and its political thematisation as embodying, or failing to sufficiently embody, a consensual etiology.

7.4.1 Economic interests and ideological factors

The first thing to note when we attempt to construe a consensual etiology of nuclear power is that need and safety have been, and still are, the pivotal concepts in establishing either support or opposition to nuclear power (Thompson 1984, p.58). In the early days of nuclear power, political consensus rested mainly on energy need interpretations that were couched in terms of economic necessity with strong utilitarian overtones.

It was argued, both in the U.S. and in Germany, that opting for nuclear power was an unparalled means to ensure the vital national objective of energy self-sufficiency and to promote economic growth (later this argument gained some additional backing by the threat of being blackmailed by arabian oil producing countries); that nuclear power was a means to meet the costs of fossil fuels that would be rapidly ascending towards the end of the century due to the depletion of natural resources; and that nuclear power was the only option available for maintaining a high standard of living as well as make it available to all segments of society.

What a minority called a "faustian bargain" was generally fairly well received as a great bounty for mankind: as a promise to "the greatest future ever spread before mankind with dazzling possibilities of life, liberty, and the pursuit of happiness" (Merriam 1947), providing virtually limitless energy that would be "too cheap to meter", as the enthusiastic phrase went (Del Sesto 1979, chap. 2).

What in the phrases just quoted is presented as embodying a morally worthy general interest was in fact never put to the test of dialogical universalisability. The utilitarian prophets helped selling nuclear power to the masses by providing an ethical rationale for pro-attitudes - they did not, however, care to ask the public or to help people to critically consider their long term interests, nor did they encourage people to opt themselves for or against risks, costs, and benefits that were determined for them by experts of economics and physics. Hence, important decisions were made on behalf of the people affected with neither the generality constraint nor the autonomous evaluation constraint being met.

In capitalist societies there is, of course, general agreement that people as contractors of goods and services are morally permitted to do as they please without critically considering their long term interests. The "free market" will gauge needs, values, wants, risks taken and products offered. It is, however, itself a morally relevant question which ranges of goods and services the people agree to be subsumed under market conditions and which not. Now, even on a very narrow economic construal of nuclear power as a promising technologically innovative product it is obvious that people's chance to express consent or dissent by means of buying or refusing to buy were unduly diminished as government goals ("national self-sufficiency") coincided with the power-industry's vested interests and led to a massive protection of these interests.[2]

In retrospect, the early pro-nuclear political consensus and public policy-making based on that consensus appear to have been in the grip of some powerful ideological motives. Powerful ideological motives internally coerce practical discourse and contribute to shortcomings in regard to the role-taking constraint, here: vis a vis future generations. Some such motives presumably were:

- The motive of embarking upon a thrilling technical challenge: a strong and ambitious faith in science and technology[3]

- The motive of turning the curse of nuclear power into a blessing: to shift the definition of nuclear energy from evil military application to something that could benefit mankind.[4]
- The belief in the unlimited capacity of human intelligence to manage self-imposed problems and, generally, a "deeply embedded belief in the utility of science for achieving practical goals" (Del Sesto 1980, p.50).

This latter belief serves to throw into relief the prominent role of scientific expertise in public policy making.

7.5 Mandated science

In a world in which science coupled to technology has become the dominant productive force, policy making and scientific expertise become inextricably entwined. We depend on scientists, e.g., to tell us whether we should be worried about radiation, whether nuclear power plants are safe, whether the greenhouse effect can be overcome, etc. It has widely been argued, not only that scientific knowledge is intrinsically valuable but also that, because it is the only truly valid type of knowledge, it necessarily leads to practical benefit (Weingart 1970). Science, so the story goes, is unique in its cumulative acquisition of unquestionable facts obtainable only so long as scientists are allowed to approach the study of nature with values that curb human tendencies towards bias, prejudice and irrationality. This selective characterisation of science, in the political context amounts to the creation of a professional ideology. Values inherent in science are described by scientists in terms, such as independence, emotional discipline, impartiality, and objectivity. For a long time, sociology of science has uncritically taken these self-interpretations at face value without questioning them as rhetoric that scientists use in order to ascertain their prominent position in society's distribution matrix of power and prestige (Mulkay 1979).

Today it seems safe to say that scientists' entry into the political arena affects them in at least three ways:

a) it influences their definition of technical problems, definitions which cannot themselves be decided by observation and systematic inference alone;

b) it influences the choice of assumptions introduced in the course of informal reasoning and the informal acts of interpretation that are essential to give meaning to "purely scientific" findings. (An example of this is the "linear" vs. the "threshold" model in the debate over permissible radiation standards.) And finally,
c) it subjects scientists to the requirement that their conclusions be politically useful. (Mulkay 1975, p.114)

The latter fact indicates a re-definition, or social role change, that scientists have to cope with once they operate as experts or expert advisors in the political context: in that context they are being perceived as purveyors of certified knowledge. They have nothing to offer other than the supposed certainties of science; and if they were to present their conclusions as no more than plausible guesses based on uncertain foundations, they would carry little political weight.

Liora Salter, a Canadian sociologist who has extensively studied the role of science in the making of standards, has coined the term "mandated science" for science used or interpreted for the purposes of making policy (Salter 1988). Mandated science transforms scientific knowledge into policy recommendations. For one part, scientific backings for policy recommendations are simply necessary for factually informed decisions to be made. On the other hand, scientific backings for policy recommendations, due to the symbolic moral capital enshrined in the scientific image, often serve to enhance trust in a preferred option and to morally discredit proponents who advocate alternative options but are unable to marshall scientific evidence in their favour.

7.6 Why mandated science cannot be a substitute of a morally relevant consensus

Let me now point out some structural features of mandated science and their moral relevance for a consensual etiology account of public policy based on mandated science.

Uncertainty
The first important feature of mandated science is uncertainty due to the fallible nature of proper empirical scientific knowledge. Whereas in non-

mandated science uncertainty is a tenet of adequate science, an assumption of its working practitioners, uncertainty causes problems for regulators and those who want to use science to support their decisions. Mandated science is thus directed towards closure, towards the production of conclusions that would support decisions taken in a nonscientific sphere of activity, whereas if scientific research is designed to be an open-ended exploration of the characteristics of natural phenomena, the result presumably will be an ever more complex and indeterminate picture of that reality.

The Rasmussen- Report on reactor safety[5] nicely illustrates this point. For proponents as well as opponents of nuclear power, this report (and its magical number of $1.7x10^{-4}$) has become a fixed point of reference in the debate over need and safety. This has tended to obscure questions of strong evaluation that are implicit in the seemingly technical terms "need" and "safety" in which the debate is framed. Why should it not turn out that nuclear power may be safe, but unnecessary? Or necessary, but unsafe? And if necessary, "necessary" in what sense? Necessary, perhaps, in order to "continue the expression of Western values of wealth, economic freedom, and opportunity through the development of industrial technology" (Thompson 1984, p.68; Henderson 1981)? By narrowly focusing on the quest for certainty, mandated science contributes to missing the autonomous evaluation constraint.

Communicative responsibility

A second soft spot of mandated science can be seen from its notorious problems with communicative responsibility, i.e. with problems inherent in the presentation of scientific findings to non-scientific audiences. "In order to maintain their credibility as scientists, participants in mandated science must ... speak as if they were speaking with other scientists. To be effective in the policy arena, however, these same scientists ... must speak with an awareness that others - whose preoccupations and interests are quite different - will use what they say to further goals that are unrelated to science" (Salters 1988, p.8) and that will probably create immense moral costs for the public. Failure to meet communicative responsibility can amount to violations of both the role taking constraint and the transparency constraints.

Again the Rasmussen-Report serves well to illustrate the point. For many laypersons with pro-nuclear attitudes, the message of the report was simply that reactor safety had been proven. Thus they confused empirically based extrapolation with a proof. An even worse misunderstanding can be traced to a confusion of the estimated probability of an event (such as a melt-down) with the frequency of such events: The report was used by some politicians to assuage anti-nuclear fears with the argument, that a melt-down would be expected to occur only after 10.000 years.

Let us note in passing that it pragmatically seems impossible for a scientist in mandated science to act both in the role of scientist and in the role of a moral advisor: in as much as he or she is perceived to express moral value commitments his or her scientific trustworthiness will wane (Salter 1988, p.193 provides an example). Therefore, the locus of moral reflection within mandated science will have to be occupied by persons who are specifically acknowledged for that role, not by the scientists themselves.

Multiplication of dissent

A third morally significant feature of mandated science is its tendency to multiply dissent that cannot be dealt with in ways prescribed by the epistemic nature of proper science. According to many studies of the use of science in external political settings such as political debates, scientific knowledge eventually does not reduce the scope of political action, but rather it becomes a resource which can be interpreted in accordance with political objectives (Mulkay 1979, p.114; Nelkin 1971; Nelkin 1975). Therefore, opposing parties in political disputes involving technical issues can usually obtain the services of reputable scientists who will provide data to buttress their policy and to undermine that of their opponents.

Within the political arena, the problem of unwanted scientific dissent-multiplication can only be solved by fiat ("enough is enough") or by strategic action (e.g. by outbuying expertise, suppressing counter-expertise, etc.), thus violating the transparency constraint, or the power neutrality constraint, or both.

7.7 Post Chernobyl

I conclude with some brief remarks about the public debate in Germany after Chernobyl. In order to meet the generality constraint of practical discourse, three obstacles to informed public discussion would have to be overcome, namely (1) lack of full public information, (2) lack of adversarial scientific expertise, (3) unwillingness of the government to foster discussion of a topic which they consider either too complex and technical for laymen or more appropriate as a subject for a public relations program to allay public anxiety. The repercussions of the Chernobyl disaster have gone a long way towards overcoming the first and second obstacle. An extensive survey of public opinion changes from 1986-1988 (Peters 1989) revealed that the majority of the German population did not shut their eyes to contradictions in the published information about the disaster and that the public was (and still is) oriented both towards statements from established institutions (government sources, nuclear power industry, nuclear research centres, etc.) and to statements from alternative institutions (ecological institutes, Green Party, Citizens' Action Committees against nuclear power, etc.).[6] "This result may be interpreted as a widespread readiness to be critical of information and a preference for using different information sources with different perspectives" (Peters 1989, p.10). However, the third obstacle (unwillingness of the government to foster critical discussion with the widest laypeople participation possible) remains. Unless such discussions can take place, neither the generality constraint nor the autonomous evaluation constraint of practical discourse will be fulliled.[7]

In the scope of this chapter I cannot discuss any substantial proposal (such as, e.g., "science courts") as to how informed public discussion about nuclear power could be made more congruent with the exigencies of practical discourse.[8] But certainly, such experiments would have to be in the spirit of John Stuart Mill who once remarked: "The only way in which a human being can make some approach to knowing the whole of a subject is by hearing what can be said about it by persons of every variety of opinion, and studying all modes in which it can be looked at by every character of mind."

Notes

1. "Discursive world" is a world in which (1) there is an exchange of speech acts between at least two persons P_1 and P_2 that we can conceptualize as an episode of argumentation (about an utterance for which universal validity claims are raised) (2) such that P_1 and P_2 mutually share an understanding of their exchange as an

episode of argumentation (3) and P_1 and P_2 could share their mutual understanding with us.

2. Technology came into being as a result of governmental investment and is growing as a consequence of governmental support. Its hazards to the health and safety of the public are not reflected in its costs because of the exculpatory effect of the Price-Anderson Act. [A parallel holds for Germany.] Since the absence of market restraints deprives the public of the opportunity to vote with its dollars on the question of risks versus benefits, the public can participate in the risk/benefit determination only through its vote at the polls. The public is entitled to this vote and to the maximum feasible articulation of the risk/benefit problem in the political arena (...) why, in a democracy, should the public not have the full opportunity to decide for itself, rationally or irrationally, what benefits it wants and what price it is willing to pay?" (Green 1970, p.137f.).

3. "Testimony was dominated by this faith in science and technology; one need only look at the masses of technical detail marshalled by pro-nuclear witnesses to get some idea of its importance. For example, AEC and industry witnesses framed their testimony almost entirely in terms of highly complex scientific and technical data. In fact, over 80 % of all testimony given by pro-nuclear witnesses was couched in terms of technical, facts-and-figures, or administrative expertise, as compared to some 17 % for anti-nuclear groups" (Del Sesto 1980, p.48).

4. A discussion of various psychoanalytic explanations of pro-nuclear consensus, cf. Kettner 1989.

5. Rasmussen Reactor Safety Study of the mid-seventies was and still is the primary evidence for proponents of nuclear power to redeem their safety claim. The Rasmussen study derives a mathematical frequency for accidents by extrapolating upon data for failures of specific components in safety systems - it is by no means a proof of reactor safety. Although Rasmussen did "a poor job of translating the method and results of the study into common sense terms"(Thompson 1984, p.66), and despite of a very critical review by the Nuclear Regulatory Commissions (Risk Assessment Review Group Reptor to the U.S. Nuclear Regulatory Commission, Washington, D.C.: USNRC, 1978) the Rasmussen study cannot be said, as some would have it, to be thoroughly disredited. "A complete formal and epistemological critique of risk assessment methodology would be required to document opposition to safety claims based upon its results" (Thompson 1984, S.66). However, the Chernobyl incident has provided so massive prima facie evidence against the safety conclusion that many people now consider Chernobyl as an informal or inductive reductio ad absurdum of any safety study, no matter how it is methdologically conducted.

6. Awareness of competing opinions and options must be regarded as rationally preferable to single minded ignorance of alternative opinions and options. The latter condition characterized media discourse and public opinion on nuclear power in the U.S. at least till 1966 (Gamson & Modigliani 1989, p.15).

7. Of the lessons that the German Parliament drew from the cognitive dissonances that riddeled the mass media after Chernobyl was to implement new regulations to the effect that information policy concerning nuclear hazards would henceforth be centralized at the Federal Government. This constitutes, in my opinion, a clear violation of the autonomous evaluation constraint (it is a paternalistic decision that sparing people some anxiety is more valuable than their unimpeded access to "all the news fit for print"). In fact, asked "whether they would prefer a more centralized or a more decentralized information policy in cases similar to the Chernobyl disaster about 62 % of the respondents ... expressed their preference for a decentralized policy" (Peters 1989, p.10).

8. Esp. Shrader-Frechette 1985, chapter 9, and p.313 for the quote by J.S.Mill. For an example of a workable framework for public deliberation about a complex bioethical issue, see Crawshaw et al. (1990).

References

Apel, K.-O. (1972)
The apriori of communication and the foundations of the Humanities. *Man and World*, vol.5, 1, p.3-37

Apel, K.-O. (1978)
The conflicts of our time and the problem of political ethics. In F.R.Dallmair (ed.): *From Contract to Community: Political Theory at the Crossroads*. N.Y.: Marcel Decker.

Apel, K.-O. (1979)
Types of rationality today, in T.Geraets (ed.): *Rationality Today*. Ottawa: Univ. Press, p.307-340.

Apel, K.-O. (1981)
Charles S. Peirce: From Pragmatism to Pragmaticism. Amherst: Univ. of Mass. Press.

Apel, K.-O. (1982)
Normative ethics and strategical rationality: The philosophical problem of a political ethics. *Graduate Faculty Journal*, vol.9, 1, p.81-108

Apel, K.-O. (1987)
The problems of philosophical foundations in light of a transcendental pragmatics of language, in K. Baymes & J. Bohman & T. McCarthy (eds.): *After Philosophy. End or Transformation*? Cambridge, Mass.: MIT Press, p.250-90.

Apel, K.-O. (1988)
Diskurs und Verantwortung. Frankfurt: Suhrkamp

Apel, K.-O.(1990)
Is the ethics of the ideal communication community a utopia? In: S. Benhabib and F. Dallmayr (1990:23-59)

Benhabib, Seyla and Dallmayr, Fred (1990),
eds: *The Communicative Ethics Controversy*. Cambridge, Mass.: MIT- Press

Crawshaw, R. & Garland, M. & Hines, B. & Anderson, B. (1990)
Developing principles for prudent health care allocation. The continuing Oregon experiment. *West. J. of Med.*, 152, p.441-446.

Del Sesto, S.L. (1979)
Science, Politics and Controversy. Boulder: Westview, 1979.

Gamson, W.A. & Modigliani, A. (1989)
Media discourse and public opinion on nuclear power: A constructionist approach. *Am. J. of Sociology*, vol.95, 1, p.1-37.

Green, H.P. (1970)
Nuclear power and the public. In M.Foreman (ed.): Nuclear power and the public. Minneapolis: Univ. of Minnesota Press.

Habermas, J. (1989)
Moral Consiousnous and Communicative Action. Cambridge, Mass. MIT Press

Habermas, J. (1990)
Discourse ethics: notes on a program of justification. In S.Benhabib & F. Dallmayr (1990: 60-110)

Henderson, H. (1981)
The Challenge of Decision Making in the Solar Age. In D.Brunner et al. (eds.); *Corporations and the Environment: How should Decisions be made?* Los Altos: William Kaufman.

Kettner, M. (1989)
Ausstrahlungen. Psychologische Argumente in der Tschernobyl-Diskussion. In H.J.Wirth (ed.): Nach Tschernobyl, Frankfurt: Fischer, p.99-122.

Kohlberg, L. (1990)
The return of stage 6: Its principle and moral point of viuew. In T.Wren (ed): *The Moral Domain. Essays in the Ongoing Discussion between Philosophy and the Social Sciences*. Cambridge, Mass.: MIT-Press (p 151- 181)

Merriam, C.E. (1947)
On the Agenda of Physics and Politics. *Am.J. of Sociology*, 53, p.167-173.

Mulkay, M. (1979)
Science and the Sociology of Knowledge. London: George Allen & Unwin.

Nelkin, D. (1971)
Scientists in an environmental controversy. *Social Studies of Science*, vol.5, p.245-61.
Nelkin, D. (1975)
The political impact of technical expertise. *Social Studies of Science*, vol.5, p.35-54.
Peters H.P. et al. (1989)
"Chernobyl" and the nuclear power issue in West German public opinion. Preprint, Kernforschungsanlage Jülich GmbH.
Salter, L. (1988)
Mandated Science: Science and Scientists in the Making of Standards. Dordrecht/Boston/London: Kluwer.
Shrader-Frechette, K.S. (1985)
Science Policy, Ethics, and Economic Methodology. Dordrecht: Reidel.
Thompson, P.B. (1984)
Need and Safety: The nuclear power debate. *Environmental Ethics*, 1, p.57-70.
Weingart, P (1970)
Die Amerikanische Wissenschaftslobby. Düsseldorf: Bertelsmann Universitätsverlag.

8 THE ARGUMENT ABOUT A NEW PARADIGM FOR HEALTH RESEARCH

Rainer Hohlfeld
University of Erlangen- Nürnberg

8.1 Introduction

Currently the elucidation and explanation of living phenomena at a molecular level - the field of molecular biology - is entering a new arena: the complex biology of human and their errors-the domain of medical enterprise. The fusion of molecular biology and medical research can be defined as 'biomedical research'. Its potential is supposed to yield enormous impact to conquer diseases so far still a scourge of mankind, like cancer, coronary diseases, allergies, immune deficiencies and hereditary handicaps. Molecular genetecists and genetic engineers can tailor bacteria or cells of higher organisms capable of producing human proteins which play a key role in metabolic regulation, growth control and development of cell lineages and organs. The most far reaching approach in the production of new drugs includes the construction and modification of the brain's special messengers, the neuropeptides, to control and master behaviourally deviating clients.

Besides all propagated benefits there is a growing uneasiness among the public as well as among younger physicians, philosophers, and scholars not affiliated with molecular biology, that health needs are not being met and that biomedical research is not having a sufficient impact in human terms. The biomedical approach - some of them are arguing - is accountable for some of the shortcomings of contemporary medical technology because the approach excludes the history of a patient‘s illness, his or her personal relationships, the doctor-patient interactions, shortly: the 'human' dimensions of disease.

In my contribution I will prove and outline the argument by proceeding in four steps:

1. I want to elaborate the current state of biomedical research as defined above, to give an example of its biotechnological potential and to conclude with the concept of 'high technology' in medicine, as it is proposed by some of the leading biomedical researchers.

2. I will provide some evidence, at least in two cases, for the shortcomings of the biomedical concepts and try to summarize the criticisms put forward against this kind of approach in a more generalizing manner.
3. I want to put the argument within the frame of reference or context of the philosophy of experimental medicine, in order to provide a fundament for overcoming the exclusive and reductionist approaches in medical research.
And 4. I want to present some leading ideas or guidelines - no more, no less - of 'inclusive concepts' in health research.

8.2 The concept of high technology in medicine

Over the last two decades biochemists and molecular geneticists developed the powerful recombinant DNA technology within the theoretical and conceptual frame of molecular biology. Through this kind of molecular technology scientists are capable of recombining genes in different developmental stages, different tissues, different species, different tribes. In other words: they are able to transcend natural barriers and to construct living things unknown to nature so far. The recombinant DNA technology is complemented by the biochemical procedures of protein-engineering and computer-aided molecular modelling, even of artificial genetic material such as the genetic instruction for a dioxin-binding protein. Both technologies mark the progress of molecular biology towards an engineering science: The goals of research are rather constructive than explorative; the objects are technical artifacts. Molecular biology enters the period of 'bioengineering' or 'constructing biology', a stage of science which is comparable to the area of chemical engineering of cyclic hydrocarbons in dye production more than a hundred years ago (1).

As I pointed out above, this concept yields the potential of an endless horizon of medically relevant human-related proteins as well as the tempting objective of constructing a 'second biological reality'. For the moment I want to address your attention to the technological power of bioengineering with respect to both methodology and ideology, which has an enormous impact in the area of medical research.

The further the molecular biologists have explored and explained the theoretical key questions of metabolism and heredity of bacteria and viruses and the molecular biology of the cell, the more they have become interested in the 'Molecular biology of Homo sapiens', as James Watson put it at the first conference for presenting the 'Apollo-project' of modern biology: the mapping and sequencing of the total human genome (2). Besides this 'biomanic' project, which is supposed to succeed within the sunset of our century, the regulation of gene expression in human cells, the processing of proteins and their precursors, the molecular mechanisms of immune response, the pathways of differentiation of skin, blood, liver muscle and nerve cells as well as the molecular mechanisms of brain function have become central issues in recent molecular biology. The underlying assumption of this kind of basic research for fighting disease is that errors in the normal functions are the key events for the molecular pathology. Through elucidation of the molecular mechanisms it should be possible to identify the biochemical, cellular or genetic defect and to compensate for it. To put it in the words of the former head of the famous Sloan Kettering Memorial Cancer Center in New York, Lewis Thomas (3): "I believe that disease results, generally, as the result of biological mistakes: misinterpretations, on the part of cells and tissues, of signals; misuse of information." And arguing for the inevitability of death, he continues: "At a certain age, it is in our nature to wear out, to come unhinged and to die, and that is that. My point here is that I very much doubt that the age at which this happens will be very drastically changed, for most of us, when we have learned more about how to control disease. The main difference will be that many of us will die in good health, in a manner of speaking, rather after the fashion of Bertrand Russell. Or we may simply dry up and blow away."

Lewis Thomas concludes his remarks on the molecular biology based medical strategies in defining their technological level: " The point to be made about this kind of technology - the real high technology of medicine - is that it comes as a result of a genuine understanding of disease mechanisms, and when it becomes available, it is relatively inexpensive, relatively simple, and relatively easy to deliver." And he exemplified his notion of medical high technology by mentioning the methods of immunization against diphtheria and various virus diseases - among them polio -, the use of antibiotics and chemotherapy for bacterial infections and the treatment of endocrinologic disorders with appropriate hormones.

By outlining the ideas and notions of Thomas I have given a more narrow definition of what is termed the 'biomedical model': the exploration, explanation and control of disease phenomena at the powerful methodological and theoretical level of molecular biology (4).

In order to confirm the intriguing idea of medical high tech I want to refer to one example from the molecular biology of neuropeptides which might be seen as a paradigm of current approaches: the animal model of a metabolic defect caused by a mutagenic event of the gene responsible for the water metabolism regulating hormone vasopressin (5). As a neuropeptide Vasopressin is synthesized in the hypothalamic brain region and has a polypeptide precursor molecule in common with the neuropeptide neurophysin. The rat used as an experimental model is unable to secrete a concentrated urea, it suffers from the disease 'diabetes insipidus'. By recombinant DNA technology and microanalytical methods of molecular biology it could be demonstrated that the mutational event in the neurophysin coding DNA region causes an irregular processing of the polypeptide precursor. This results in deficiency of the functional vasopressin. In this case the advice for cure is to substitute the missing hormone. This case provides a clear example of the explanatory power of the causal approach at the level of molecular genetics and hormone biochemistry.

8.3 Halfway technology in medicine and the shortcomings of the biomedical approach

L.Thomas has defined a second level of technology, which he has termed the 'halfway technology' in medicine. Based on trial and error empirism, it is a technology " designed to make up for disease or to postpone death". The treatment of cancers, the transplantation of hearts, kidneys, liver and the invention of artificial organs for him represents this level of medical technology. Can this level be transduced to medical high technology or might there be general limits to this kind of reduction?

To follow the question let me refer to the case of the brain's own messengers, or more generally, the body's own proteins. As it has been experimentally demonstrated, human peptides or transmitters intervene in a complex pattern of interaction of the hormone, immune and nervous system. They can induce cascades of biological processes, have more than one effect or function and are released very specifically, temporarily and spatially. So far there is no experimental model providing for this complexity and the physician must take into account serious side effects due to the multiple nature of the protein messenger (6). Thus the molecular biology of messenger or drug - receptor interaction in the test tube represents one side of the coin; the internal milieu of the body, influenced by biological, psychological and environmental factors -the subject of pharmacocinetics and clinical research - is the other side.

To give an even more obvious example: through the identification of a genetic marker linked to the distribution pattern of manic depression in Amish people scientists believe to have removed the stigma surrounding such patients. "They suffer from a disease, the causes of which are beyond their control", commented the American geneticist J. Egeland (7). Therefore they cannot be made accountable for it. Besides the question of the scientific serosity of reducing the complex pattern of manic depression to the molecular biology of gene action, their interpretation is misleading: Labelling a deviating behaviour as biologically determined has served as a base for drug control in the history of psychogenetics and psychiatric diseases. The exclusion of the psychosocial dimensions of the patient's suffering has become part of the labelling and discriminating procedure. The biomedical approach is not sensitive to this kind of problem, nevertheless central for patient care in most cases.

To summarize my first critical remark on the biomedical approach: the area of halfway technology in medicine includes at least some cases which cannot be reduced to causal explanation at a molecular level. The biomedical frame of reference provides a necessary, but not sufficient condition for complex psychic disorders.

By this notion I reconfirm an old criticism of biomedicine presented more than a decade ago by the psychiatrist and medical scientist George L. Engel. He elaborated on his critics with respect to the perception of diabetes and schizophrenia as diseases - two cases which might be seen in analogy to the cases I have discussed above. He outlined: "In the biomedical model, demonstration of the specific biochemical deviation is generally regarded as a specific diagnostic criterion for the disease. Yet in terms of the human experience of illness, laboratory documentation may only indicate disease potential, not the actuality of disease in time. The abnormality may be present , yet the patient not be ill. Thus... the presence of the biochemical defect of diabetes or schizophrenia constitutes but one factor among many, the complex interaction of which ultimately may culminate in active disease or manifest illness. Nor can the biochemical defect be made to account for all the illness, for full understanding requires additional concepts and frames of reference."(8) Thus the perception of the phenomenon of illness requires "consideration of psychological, social and cultural factors." Evaluating the dominance of the biomedical model in medicine he remarks: "The biomedical model has thus become a cultural imperative, its limitation easily overlooked. In brief, it has now required the status of a dogma....Biomedical dogma requires that all disease, including 'mental' disease, be conceptualized in terms of underlying physical mechanisms."

8.4 The world of the medical and biomedical model

The reduction of complex illness patterns to underlying molecular mechanisms, the reduction of behavioral disorders and melancholy to brain chemistry do not result from progress in medical research. They are expression of the philosophy of experimental medicine, which applied the experimental methods of the physical and chemical sciences to biology and medicine during the 18th and 19th Century.(9) This philosophy laid the foundation for the paradigm and selfinterpretation of scientific medicine: the medical model. According to it, the body is a machine and the disease is nothing but the breakdown of the machine, and the doctor's task is the repair of the machine. Thus, the scientific approach to disease began by focusing on physio-chemical parameters of physiological processes and excluding the behaviourial and the psychosocial contexts of diseases. The fractionation of the medical disciplines can be seen as a consequence of the experimental division of labour in exploring and explaining the human body. Through the transformation of the medical into the biomedical model as a consequence of replacement of traditional theories in biology by molecular biology the technological and control power improved remarkably; but the underlying philosophy did not change in any way.

The focus on the biochemical or genetic lesion has the effect that reality perception and construction through this scientific approach lack the personal, social, environmental and historical dimensions of the patient's illness. The notion of human as 'being a subject' disappears under the cultural command of the biomedical dogma. If the 'Molecular biology of Homo sapiens' proceeds further, then within the picture of the world of biomedicine problems of living, conflicts, grief, love, moral will become nothing but a highly complex arrangement of molecules, sometimes out of order, but gifted with the natural power of self organizing new patterns of order and function. And if we can understand all parts of the living machinery we can even try to 'optimize' or even 'enhance' the most vulnerable and poorly adapted parts. The American geneticist Bentley Glass have put this dream of biomedical reasoning into the words: "As man acquires more fully the power to control his own genotype and to direct the course of his own evolution, he must produce a Man who can transcend his present nature."(10)

8.5 Prerequisites for an inclusive model in health research

Perhaps with the power and dominance of the biomedical model as a 'cultural imperative' the talk of inclusive approaches in the world of medicine currently is nothing but the summarizing of shortcomings, the definition of some prerequisites and the giving of some guidelines for the search for an inclusive model in health research. In defining the prerequisites I again refer to G. Engel: "To provide a basis for understanding the determinants of disease and arriving at rational treatments and patterns of health care, a medical model must take into account the patient, the social context in which he lives, and the complementary system devised by society to deal with the disruptive effects of illness, that is, the physicians role and the health care system. This requires a biopsychosocial model."

How can this word be more than a phrase? Engel himself referred to the holistic approach of Bertalanffy's general systems theory (11). But this might not be sufficient. If the holistic approach does mean the integration of elements, functions and structures of a living 'system', the approach may overcome the almost extreme monocausal reductionist interpretations (12), but it still sticks to the machine theory of living phenomena put forward by the most advanced physiologists in the 19th Century like Claude Bernard. If the general systems approach should provide guidelines for conceiving machine-like systems only the main task for the inclusive approach is not fulfilled. That is the question of the relations and interactions between the physiological, psychic and social dimensions of illness. That is the question of inclusion, not only of elements or subsystems, but even of different frames of references or cognitive contexts such as molecular biology, behaviourism, psychology, psychosomatic medicine, sociology and epidemiology (13).

References

1. Winnacker, E.L. Synthetische Biologie, in: Herbig J, Hohlfeld R (eds): *Die zweite Schöpfung*, München, Hanser 1990, pp 369-385.
2. Lewin, R. Molecular Biology of Homo sapiens. *Science* 1986;232: 157-158.
3. Thomas, L. *Aspects of Biomedical Policy*, Washington, Nat. Acad. of Sciences 1972.
4. Hohlfeld, R. Das biomedizinische Modell, in Herbig J (ed): *Biotechnik*, Reinbek, Rowohlt 1981, pp 114-134.
5. Richter D, Ivell, R. Gene Organisation, Biosynthesis and Chemistry of Neurophyseal Hormones, in Imura, H. (ed) *The Pituitary Gland*. New York, Raven Press 1985,p 127.
6. Scriba, P.C, Werder, K. *Das Konzept der neuroendokrinen Regulation und seine therapeutischen Konsequenzen*. Verhandlungen der Gesellschaft deutscher Naturforscher und Ärzte 1987; 114:223-240.
7. Kolata, G. Manic Depression Gene Tied to Chromosome 11. *Science* 1987; 235:1139-1140.
8. Engel, G.L. The Need for a New Medical Model: A Challenge for Biomedicine. *Science* 1977; 196:129-136.
9. Bernard, C. *An Introduction to the Study of Experimental Medicine*. New York, Dover 1957.
10. Glass, B. Endless Horizons or Golden Age. *Science* 1971; 171: 27-29.
11. Bertalanffy. L. *General System Theory*. New York, George Braziller 1968.
12. Polanyi, M. Life's Irreducible Structure. *Science* 1968; 160:1308-1312.
13. Uexküll, T., Wesiack, W. *Theorie der Humanmedizin*. München, Urban & Schwarzenberg, 1988.

9 AIDS AND HUMAN RIGHTS: A SOCIETAL CHOICE Juridical reflections on the spread of H.I.V.

Daniel BORRILLO
GERSULP (Groupe d'Etude et de Recherche de la Science de l'Université Louis Pasteur) Strasbourg

9.1 Introduction

It was not so long ago that the industrialized societies thought that they had conquered the plagues of infectious diseases. Mortality rates in France showed that the principal causes of death had become: heart disease, cancer, suicide and traffic accidents[1].

In 1981, however, United States federal authorities noted a considerable increase in the distribution of a medicine named *pentamidine isethionate*, used in the treatment of a type of pneumonia known as pneumocystis carinii, which affects sick patients whose immune system had been deteriorated by cancer. The illness was so unusual to the extent that its treatment was considered purely experimental. Between 1967 and 1979 there had been only two prescriptions written for this previously mentioned medicine.

The first five cases of AIDS discovered in 1981 had several points in common. They all involved young homosexuals whose immune systems, up till their diagnosis, had been functioning normally. At the same time, the appearance of several cases of a known but quite rare cancer named Kaposi's sarcoma and pneumonia was discovered in the same population cited above. Other equally peculiar illnesses were progressively detected which had the same common denominator, that being the same impairing effect on the immune system.

The relationship was quickly established and in 1982 was recognized by the clinical community. The newly identified syndrome was first called G.R.I.D. - Gay Related Immune Deficiency (whose terminology was quickly considered too limited because it makes a relational link between an immune deficiency and male homosexuality), and later called A.I.D.S. - Acquired Immune Deficiency Syndrome - which was considered more appropriate. AIDS is caused by a chronic immune deficit whose etiological agent is a retrovirus[2] which represses the most serious infection of the virus. The illness was detected in the same year (1982) in a group of several Haitian and hemophiliacs. This confirmed the hypothesis that the transmission of the virus was through both blood and sexual activities. In 1990, the World Health Organization (W.H.O.) estimated that there was approximately 283,000 full blown cases of AIDS and that more than 8,000,000 individuals who are carriers ofan asymptomatic form of the virus. These individuals are known as 'seropositives.' Faced with this peculiar illness, the international press' first reaction was to establish a direct relationship between the disease and the "gay community." In the eyes of the general public, the illness was presented as a sort of "Homosexual Chernobyl."[3]

The history of epidemics well illustrate how man has reacted in the face of a plague from an unknown origin. In trying to find their "evil" origins in certain social groups, as in the case of the "Black Death" plague of the 8th Century which decimated Europe, the Islamic nations played the role of 'onlookers', absolved from the drama. Their first reaction was to identify the "evil" with the Christian communities who generally appeared the most affected (probably because they stayed closely together in tight circles). To think that a virus could choose its victim according to his professed religion would seem absurd today!

In the same fashion during the Second Plague (1348-1352) which in no time whatsoever wiped out a third of Europe's population, Christianity considered it a punishment from God, whose divine anger found the responsible parties. The Catholic Church found its scapegoat in all who didn't profess to the Christian faith. It wouldn't be necessary to recall the expulsions of the Jews and the Gypsies from major European cities during this period.

Man has always found himself needing to explain the catastrophes to which he is regularly subjected. For this matter, we are sometimes surprised by the reasoning used to analyze and to represent epidemics[4]. Apart from the previously cited examples, we are able to furnish an interminable list of interpretations given to illnesses. Let's take several examples: Various ancient Mesopotamian texts provide us with explanations of illness brought on by the anger, and even caprices of the gods. Freckles appearing on a baby's skin as a consequence of an infectious disease were interpreted as scratch marks of the devil *Iamashtu*[5]. Later, the concept of "moral conscience" was born. The gods no longer pursued innocents, but punished the guilty (those who violated sacred places, who did not respect agricultural rites or who committed incest, notably).

Though the Occident had undergone a process of secularization over the years, the causal relationship of disease-responsible/disease-punishment had not altogether disappeared from modern medicine, e.g. the victim of lung cancer is punished for having smoked too much, hence those with AIDS are being punished for their loose sexuality or promiscuity[6].

Over the years, the relationship of AIDS-homosexuality fulfilled a reassuring function for the rest of the "heterosexual society." A "normal" sex life permitted people to feel apart from the scourge of AIDS. It wasn't until 1986 that AIDS became an illness which came to concern everybody by the simple fact that the possibility of a heterosexual transmission was proven.

According to M. Pollak[7], AIDS, having its origin in the closest held taboos of Occidental society - blood, sperm, sex, homosexuality and death, has created a quasi-experimental situation by putting to a test the meanings of tolerance and freedom. It thereby challenges the capacity of a democracy to respond to unforseen threats to the very pillars of its society.

9.2 Epidemic Management by the rule of law

A democratic society characterizes itself principally by its government's organization, structure and its responsibility to its citizenry and the individual. Constitutional law is characterized by the respect of what we call the rule of law[8], formed around two major principles: *the principle of legality* which is a complex ensemble of formalities and procedures whose goal is the distribution and limitations of powers which protect personal freedoms. The second is *the democratic principle* which, because the constitutional state must be democratic, as it grants the population access to power.

The history of infectious disease management and epidemics is characterized by strict measures of control and by limiting personal liberties in the name of the general health risks to the society at large[9]. Beyond the generally accepted reasons for reducing the possibility of a contagion, the management of an epidemic responds to the structural exigencies which embrace a larger vision...that of the political agenda behind the government's epidemic management.

We find two principal tendencies in the following public health propositions. The first tendency is to take into account the premise of the constitutional state and the sick citizen, and the other of a strong policy of interventionist management based on a type of integrationist ideology which justifies a type of defense of the "seronegative society."

Freedom is in danger when the State assumes the power to decide in the place of the individual, for liberty by definition limits the State to that which is necessary to protect the rights of others. This jumbled ensemble of rights which explain the notion of public order is where we find the category of "public health." The existence of these two restrictions inevitably opens to unending debate the sensible limitations which are capable of bearing on our personal freedoms. Professor Patrick Wachsmann noted that the European Convention on human rights carries a precious clarity when it specifies that the restrictions of certain liberties which are not proclaimed legitimate are conditionally subjected to being "necessary in a democratic society for the safeguarding of certain values which are the rights of others and the protection of public health.[10]"

Which brings us to think about two major and fundamental principles in the management of the epidemic and the rule of law: the refusal to adopt emergency legislation and the prohibition of the State to interfere in the personal decisions of its citizens. In order to insure the maintenance of public order, and asserting the existence of a crisis which puts the entire society in danger, the State is sometimes tempted to create and apply emergency legislation. Legislation which involves an extension of executive branch powers, as was the case, for example, in the September 9, 1986 French government's special directive against terrorism.

Up to now, the adoption of emergency legislation concerning AIDS has been rejected, and it is hoped that this decision should be firmly supported. Because of its means of transmission, AIDS differs from other epidemics and does not, therefore, threaten the population as a whole. Contamination is principally produced by contact with blood or sperm and can be avoided by using sterilized syringes and condoms. In the most commonly known high risk situations, e.g. blood transfusions, organ donorship or artificial insemination, an AIDS-blood test is mandatory[11]. A call to individual responsibility in prevention must be stressed. In France, the government, working in partnership with certain associations has proven itself very efficient in this effort.

9.3 The role of the physician under the rule of law

Beyond the ethical rules which doctors have come to respect as a function of their professional codes, constitutional law has established a certain number of both general and mandatory measures. Just as a medical secret comes to be represented as part of medical ethics, it is necessary to underline that it is also legally bound by Article 378 of the French Penal Code[12] which established sanctions against those who divulged confidential information gathered under the doctor-patient relationship[13]. There can obviously be no medical secrets between the doctor and the patient who has the right to his medical data. In this regard, the French National Consultative Committee on Ethics declared the necessity of informing patients, under acceptable circumstances, of their medical condition taking into account the psychological state of each patient[14]. Under French law, Decree N° 86-770 makes it mandatory to report each case of AIDS (though not the seropositives) and it must be done anonymously, to the DASS (*Direction des Actions Sanitaire et Sociale).* This mandatory reporting constitutes the sole exception to the principle of medical secrecy.

The problem becomes more complex at the moment when this sensitive medical information enters medical information systems. *La Commission Nationale de l'informatique et des libertés* (CNIL), concerned by the problems posed by the development of data access and the individual's right to privacy, adopted strict measures regarding the treatment of AIDS research and patient files[15]. Among the problems cited was the possible misuse of medical data, whether exploited for political or financial ends, which could put fundamental freedoms in jeopardy[16].

If the patient's name is required to be recorded[17] in all epidemiological cases, the AIDS patient's name must be rendered anonymous prior to his case information being placed into a medical data bank[18]. Anonymity is at the core of the reconciliation of the patient's right to privacy, and the protection of research data. Every part of the patient's file must be declared to the CNIL which examines the adequacy and relevance of the records with respect to the defined limits set forth by the declarant. All recorded information must rely on the patient's consent, who, in any case, is authorized to have access to his file but who can also reconsider his decision[19].

But this isn't the only level of the violation of legal norms put into question by the rule of law. In other widespread, though less serious situations, we see a profile of a distinct abuse of power on the part of the doctor, notably when he assumes a moralizing role with the patient. An inevitable dependency relationship becomes established between the ill patient and the doctor. This relationship must be accompanied by an ethical responsibility which constantly stresses that the ill patient (or the seropositive) is a private citizen - and that the doctor is in no way authorized to exploit his position and act as a moral figure in the patient's private life.

9.4 Words and Deeds in official epidemic management

Using metaphors borrowed from medical vocabulary, the ideology of hygiene has claimed to explain the functioning (or disfunctioning) of society. Such was the case when articulating the 19th and 20th Century character of Italian anthropological studies of criminality[20] which started the acclimation of certain terminology into continental European themes. These themes went on to abundantly serve totalitarian regimes who protected the "healthy body of society" from "foreign bodies" and the "virus" which came from the "infected": born criminals, congenital idiots, the insane and sexual deviants...[21]" In the authoritarian view, democracy isn't capable of maintaining the "healthy" social body and at the same time eliminating infectious elements from it[22]. Democracy is put into question, according to them, because of its laxity and dangerous permissiveness vis à vis its people and it is that which is at the origin of decadence. Too much freedom, too much tolerance towards "fags", too much understanding for drug addicts, too much contact with strangers (especially Africans). Here we have several mythical explanations and imagery which has found its way into our common parlance, spawned by certain politicians and certain members of the scientific community[23].

Metaphorical language[24] doesn't take the facts into account, it doesn't leave itself to the realization of its own prophesies. Let us remind ourselves' of the altogether unjustifiable excesses, from the scientific point of view, which we've witnessed in recent times. The closing of establishments where AIDS was spread (Decision of the Bavarian Ministry of the Interior, February 25, 1987), the expulsion of all foreigners who constitute an AIDS -related risk (ditto, May 19, 1987); the authorization to quarantine all HIV positive patients in the State of Texas (USA), the obligatory hunting down and systematic imprisonment of seropositive individuals in Cuba; the prohibition to enter American territory for seropositive and AIDS-infected individuals at the time of the most recent international AIDS conference in San Francisco; the power given to Rumanian doctors to mandatorily hospitalize infected individuals. These examples, among many others[25], are instead the result of the fruit of metaphors[26] and discursive ensembles which have strongly conditioned health policies. For ages we've heard talk of divine plagues, of a combat against a virus, of excessive freedoms and the creation of a virus which serves the purposes of conservative and puritanical forces, etc.

It is time to break away from these stereotypes and metaphors and meet the obligation of creating the most realistic and reasoned management of the epidemic. According to Susan Sontag, the most honest attitude we could have towards the illness and the most honest way of being sick oneself is to weed out its metaphor and resist the contamination which accompanies it[27].

9.5 The epidemic in the state of European law

The Council of Europe is particularly concerned with the struggle against the epidemic because, as underlined by Mme Massarelli[28], the Council's director of heath services, what is in question poses a major challenge to public health policy itself. The Council's work is carried out within a general framework of recommendations by WHO, and its goal is that its findings would influence a harmonization of European legislation on the issue. Making human rights an imperative of public health is the fundamental preoccupation and primary axis on which all of the Counsel's recommendations are organized.

In order to limit the spread of AIDS, the Council proposes to win the confidence of HIV infected persons. Any measure of coercion or discrimination could provoke a negative reaction in the infected community, as well as in the general population. It is a given that the indispensable means in the management of an epidemic is information and prevention, but it is difficult to see how a collaboration between the State, the AIDS carriers and the seropositives would come about in this management. The only strategy considered acceptable by European juridical order, with respect for both efficiency and personal rights, was that of a prevention policy composed of information dissemination, education, voluntary testing, counseling and complete respect for confidentiality. A draft bill at the Counsel of Europe was quickly introduced.

In 1983, even before the virus was isolated, the Council presented its first report which recommended that transfusion services suggest a self-exclusion policy for would be donors belonging to risk groups[29]. The French National Assembly equally demanded that the private lives' of individuals should be respected and that information campaigns should not be directed towards any one social group[30]. In 1985, once the virus was discovered, the Committee of Ministers recommended mandatory AIDS-testing of blood donors to track down the virus[31]. Other recommendations were proposed by the Counsel trying to systematically take into account the legal principles necessary in respecting individual rights at the benefit of public health[32].

9.6 Examples of loopholes in the rule of law

Despite the international institution's efforts, a number of dubious legal situations continue to arise. Beyond the serious steps taken by public authorities in a considerable number of countries, the situation in Europe is far from being worked out, aside from the cited cases of Bavaria, Belgium or Ireland - certain practices of some French institutions make a call for reflection. As an example, we'll take the case of the right to insurance and labor laws[33].

The French National Committee on AIDS recommended in its February 20, 1990 declaration that it was "forbidden for insurance companies to subordinate to conclusion an insurance policy based on the results of a blood test for the AIDS virus." It equally advised that it would "be vigilant that insurers did not introduce questions in their applications which made reference, in an explicit or indirect way, to the applicant's style of life or sexuality." Despite these recommendations, insurance companies maintain discriminatory practices to this day.

The French insurers complain that they bear the cost of medical secrets in France, which hinders the creation and exploitation of samples, which lead to actuarial tables which aid them in better ascertaining their insurance risks[34]. Mr. Pierre-Denis Champvillard, the director general of Scor-Vie, a company specializing in reinsurance confirms: "If we decide to mutualize the 'AIDS risk', it would inevitably raise the premiums for clients who are not seropositive. They'll (non-seropositive clients) go elsewhere, to London, for instance."

A draft bill which would nullify the application of Article 416 of the French Penal Code[35], the very code which protects applicant from being discriminated against by reason of physical condition or handicaps, has as its goal to allow insurers to limit their types of coverage to AIDS patients[36]. The insurance lobby is claiming that not only should mandatory AIDS-testing take place for applicants, independent of a covered risk[37], but also for the establishment of insurance industry files on all seropositives. As Pierre Lascoumes points out, the insurance companies are in the process, in effect, of obtaining the authorization to accomplish that which they swore they would never do...of knowingly excluding seropositives from individual insurance policies. Let us call attention to the fact that such a measure has never taken place for any other illness, even those which result in more deaths than AIDS. There has never been, for example, a similar requirement for mandatory tumor testings for cancer.

The situation with labor laws has been no less dramatic. The French National Committee on Ethics in its December 16, 1988 decree (2.D.) established that "being seropositive would not be considered an obstacle in the exercise of professional activities, public or private, and did not entail, for example, a job disqualification," on this general principle the committee quickly added "It might appear in the future that the exercise of certain professions will be incompatible with being seropositive - this is for two reasons - on the grounds of the transmission risk of the illness to others, and on the grounds that the pathological consequences for others may arise, as well. These situations, in all likelihood, are exceptional and should be the subject of study and special decisions."

In the same sense, it was recently recalled that relative questions of health were part and parcel of the individual's private life and by consequence deserved protection. It is on this basis that a private company was ordered to pay out damages with interest to one of its employees for having approved of the posting of a memo relevant to the seropositive state of the employee[38]. The theory is clear and simple, less so than in practice. Such is the current state of labor laws, and in considering the instruments at the employer's disposal (they are not all as naive as in Burke France), the seropositive, and more so the carrier, finds himself in a very precarious situation in the workplace. The afflicted worker progressively passes through the asymptomatic state towards the first symptoms of the illness when he is finally obliged to take leave and find and commit himself to a medical care facility. If the absences are prolonged, the employer can legitimately lay the worker off. Furthermore, no provision of the French Labor Code permits an employee to ask to rearrange his working hours in order to enroll in a medical treatment program, when in Article L. 212-4-12 of the same Code legitimizes this kind of request "for the regular and controlled practice of sport." One can say, as Pierre Le Cohu commented, that "the more you have, the more you receive; we give time off to do sports, but when it comes to a sick employee, he takes care of himself...if he can[39]."

These examples show that, with respect to law and society, the problems posed by AIDS are far from being resolved. From all the discussions by "specialists" we get the impression that the real debate has never taken place, which is, so to speak, how the thousands of seropositives, who continue to have a normal, daily existence want to work, want to play an active role in society, want to love and to live the rest of their lives' in the best way possible.

9.7 Conclusion

In this century of unending medical progress, there is an illness which takes on epidemic proportions, shocks worldwide opinion and puts the foundations of public health policies into question. The stakes are high, for the reactions and from the conflict sometimes awakens irrational, exaggerated fears of the illness. Statistically speaking, AIDS is much less widespread that other illnesses or accidental death, but it seems to have appeared in our society as a form of punishment.

For centuries, man has associated crises with the loosening of morals: The historian Sallustre estimates that in the 1st Century BC that the decadence of the time was due to the loss of *virtus* (courage, power of the soul..). Zosime, in the 5th Century AD holds the Christian religion responsible for the crisis of submissiveness leading to the loss of the healing virtue. Or more recently, in 1860, Francon writes that "If the inhabitants of the Occident are idiots incapable of withstanding hardship, then it is because they're degenerate[40]."

The philosophy of human rights gives us the possibility to think otherwise. It is in this way that AIDS reflects one of the most profound problems of our civilization, and it depends on us to place the debate on this level. To what degree is homosexuality tolerated in our society? When are we going to have an international consensus on the epidemic? What leads so many young people to turn to drugs to escape reality? Why have we chosen to attack AIDS victims in the place of the fight against the virus itself?

And so we've seen how the AIDS epidemic has put democratic societies face to face with a decisive choice; one which reveals the durability of the full meanings of tolerance, of the respect for privacy and of solidarity which we claim so much to defend. The way we manage this epidemic will surely define the society we leave to our children. In choosing the imperative for the rule of law in resolving the AIDS question, we are choosing our own freedom.

Notes

1. In 1985, INSERM (Institut Nationale pour la Sante et la Recherche medicale) printed the following statistics:
   -app. 10,000 deaths attributed to traffic accidents (9,985)
   -app. 12,000 deaths attributed to suicide (12,363)
   -app. 15,000 deaths attributed to alcoolisme (15,269)
   -app. 29,000 deaths attributed to tabacco- related illness (28,550)

2. It wasn't until after 1978 that scientists were able conceptualize and detect, due to biotechnical advances, a pathogenetic human retrovirus (H.I.V.). This data reinforces the hypothesis that the virus had existed for a long time except in a weaker less identifiable state.

3. The first headlines from the French press made exclusive reference to the homosexual community. The well known French journalist Escoffier-Lambiotte, wrote an article entitled "Mistérieux cancer chez les homosexuals américains."("A mysterious American homosexual cancer") which appeared in *Le Monde de la Médecine* of January 27, 1982. Other publications used at the time the same terminology, as

in "Lépidemie du cancer gay", *Libération* of March 19, 1983 or "Les homosexuels punis par le cancer," *Le Matin* of January 2, 1982.

4. "The history of syphilis in this respect is exemplary: the brutal appearance of this new plague, in 1493, was a mystery. Despite several discussions on the possible ways of transmission by the "miasmisms" or by breathing it in, venereal contagiousness has been recognized as its principal mode of transmission. From this mystery and this certitude should be born the belief that a divinely created disease was created to punish sinners." *Sid'aventure*, Ed. Syllepse, Paris, 1989, p. 58.

5. Ruffié J. and Sournia J.C., *Les épidémies dans l'histoire de l'homme*, Flammarion, Paris, 1984.

6. Sontag, S., *La maladie comme métaphore*, Le Seuil, Paris, 1979.

7. M. Pollak, *Les homosexuels et le sida: sociologie d'une épidémie*, Métaillé, Paris,1988.

8. The Rule of law is the rule which must impose itself on all citizens, rich or poor, the governors or the governed. It is also known as a constitutional state or a democratic republic.

9. "Faced with leprecy, exile and detainment were organized to separate lepers and non-lepers. The plague was accompanied by various sectors which kept urban populations in their city, in their towns, in their districts and in their houses. The inhabitants were summoned to their windows so as to count the living, the sick and the dead." Defert, D., "Epidemics and democracy", *Actions et Recherches Sociales* N° 3, September 1988, P.33.

10. Wachsmann, Patrick, "AIDS or the management by fear by the State of law" in *Sida et droits de l'Homme: l'épidémie dans un Etat de droit,* text from a conference convened by Borrillo D. and Masseran A., Actes du Colloque, Gersulp, Strasbourg, 1990.

11. Memorandum of the French *Direction Générale de la Santé* October 20, 1985 and October 28, 1987.

12. Professional secrets are bound under this code such that "all persons endowed by either the state or professional codes, or by a temporary or permanent function to whom secrets are confided" (religious, notary, process servers, etc.).

13. It would be difficult in this paper to thoroughly examine the French judicial system and the question of medical secrecy. We can refer interested readers to the work done by D. Thouvenin in *Le secret médical et l'information du malade*, Presses Universitaires de Lyon, 1982.

14. Notice of the Comité Consultatif National d'Ethique for science, life and health for the problems posed by the fight against the spread of infection from HIV (Human Immune Deficiency), December 16, 1988.

15. See E. Heilmann, "Medical Information Systems and the protection of nominative data in respect to HIV," *Actes - Les cahiers d'action juridique)* N° 71-72, June 1990, P. 36.

16. In a press release of November 1989, *L'association Aides* denounced the creation of a seropositive file by various insurance companies

17. Deliberation of the CNIL of September 1988 (Assistance Publique de Paris) 9° Report of Activity, 1989, pp. 344-345.

18. Deliberation of the CNIL of July 5, 1988 (Assistance Publique of Marseille) 9° Report of Activity, 1989, pp. 338-340.

19. The French Law entitled "Information technology and personal freedoms," January 6, 1978.

20. Especially with Lambroso and Ferri, In France, this school is principally represented by Lacassagne and Gabriel Tarde.

21. See Patrick Tort, "On cases of pscyho-sociological and fascistic rhetoric in the discussion of health ." *Sid'aventure*, op. cit. p. 29.

22. In France, Le Front National claims that human rights legislation impedes an efficient approach to fighting the AIDS epidemic.

23. To the extent that Dr. Bachelot publically declared that AIDS-infected persons are "veritable bacteriological bombs" necessitating the creations of "aidsatoriums" in which to confine "aiddicts" who risk contaminating the healthy majority of the population. Interview in *Libération*, February 1987. Not to forget the declarations of MR. L. Pauwels in *Le Figaro magazine* of December 6, 1986: " A group of measures for the society for not having taken disappear: selection, the promotion of personal efforts and individual responsibility, national codes, the fight against drug use, etc. the heresies. This return to reality is a scandal for them. They are afraid of their lack of morals. Here is their revolutionary sentiment. What we have here in our youth is a mental AIDS. They've lost their natural immunity; every terrible virus is attacking them."

24. Aristotle wrote that metaphor consisted of giving something a name which belonged to something else.

25. In Great Britain, there is a draft bill (Criminal Justice Bill, Article 25) which would re-criminalize homosexuality. In Belgium, foreign students from non-EEC member countries are required to present a certificate of non-seropositivity at the time of enrollment. In Ireland, seropositive inmates are placed in special cells and subjected to "solitary confinement." Despite the Counsel of Europe's recommendations, Ireland and Belgium have not accepted to implement any promotional campaign on condom usage.

26. According to Sontag, AIDS has a double genealogical metaphor. Insofar as a microprocess goes, it is described like cancer: it's an invasion. When we throw in the mode of transmission, we run into an even older metaphor , linked to syphilis: pollution. (It is contracted through blood, from sexual fluids of an infected person or by contaminated blood products). *Le sida et ses métaphores*, Ch. Bourgois, Paris, 1989, p. 24.

27. Sontag, S., *op. cit.* p. 9.

28. Botho-Massarelli, Vera, "Ethical repercussions of AIDS in the framework of health and social policies," *Sida et droits de l'homme: l'épidémie dans un Etat de droit,* op. cit., p. 19.

29. Recommendation N° 8 on transmission prevention possibilities of AIDS acquired by contaminated blood donation or from blood products.

30. Resolution 812 (83) relative to AIDS

31. Recommendation 12 (85) on AIDS-testing for the presence of the signs of AIDS in blood donors.

32. Recommendation 12 (85) in relation to a common European health policy against the spread of AIDS. Recommendation 1080 (88) of the Parliamentary Assembly. 8th Conference of the national prison administrators. Recommendation (89) on the ethical implications within the health and social policies. Recommendation 1116 (89) of the Parliamentary Assembly concerning AIDS and human rights.

33. In other disturbing situations taken from some non-democratic positions can be found in common law, criminal law, penal law, etc. Interested readers will find more thorough explanations in N° 71 of *Actes* "AIDS and the law," June 1990. From the records of a colloquium held at GERSULP, *Sida et droit: la régulation juridique d'une épidémie*, Arles, Ed. Actes Sud, to be published.

34. Jean-François Rouge, "The Economics of AIDS," *L'expansion*, Jan. 24 / Feb. 6, 1991, p. 54.

35. Draft bill N° 1182 (1990).

36. See Lascoumes, Pierre, "When Insurance companies hunt bad risks. Seropositivity, an overestimated risk," *Actes*, op. cit., p. 72.

37. For seropositives having today, on average, eight more years of normal life before becoming full blown, the main problem isn't life insurance, but that of insurance linked to professional loans or supplementary insurance. The approbation of this draft bill involves an inadmissable discrimination.

38. The "Burke France" affair, Paris Grand Jury decision of June 7, 1989, cited by Françoise Degott-Keiffer, "New illnesses and the right to work, the case of HIV infection" in *Sida et droit de l'homme: l'épidémie dans un Etat de droit,* op. cit. p. 138.

39. "Employment, labor laws and AIDS," *Actes,* op. cit., p. 45.

40. These examples are analysed by Pascal Hintermeyer, "The AIDS idea" *Action et recherches sociales,* op. cit pp. 69-70.

PART V

Public and State Interests in the Development and Control of Technology

10 BIOTECHNOLOGY AND SOCIAL PERCEPTION

Olaf Diettrich,
Commission of the European Communities, Brussels

10.1 Introduction

Everybody knows the fairy-tale "Beauty and the Beast" of the innocent maiden and the ugly and horrible beast which turned out to be something honest and trustworthy after being treated with goodwill and trust rather than with fear and repulsion. Intuitively one may think that this is a somewhat optimistic but nevertheless appropriate metaphor to describe the complex and tense relation between the public and biotechnology. But, unfortunately, there is no evidence for who is the beauty and who the beast. Is it the innocent and trusting public which is confronted with a pullulating science threatening human life, the environment and the integrity of God's creation? Or is it a pure and beneficial science promising progress in nearly all human problems which is rejected by an ignorant and distrustful public?

The Commission of the European Communities which (like many public authorities) has an explicit mandate (or decided explicitly) to promote the social welfare and the economic prosperity of the public as well as to protect the environment and the diversity of terrestrial life, to develop research and science and to foster the competitiveness of European industry, can hardly define priorities between these goals as those usually do who have to represent mainly one single position. Environmental groups for example may say that a functioning environment is the prerequisite of any life on earth and therefore requires first priority.

Industrial associations will argue (1) that "Biotechnologies promise new opportunities for economic growth, new job creation, industrial renewal, environmental management and revitalised strength in the agricultural merket place. Future European competitiveness on a par with the U.S. and Japan in the many industries which will depend on biotechnology must therefore become the principal objective of Community policy". Science considers itself as the very producer of any progress and, therefore, must be the focus of all public policy. Ethical positions, last but not least, take by definition precedence over any other kind of arguing. So the environmentalist may tend to load an economic burden upon society which the social politician will hardly agree to pay. Or the scientist who wants to remain in the favour of public appreciation would try to tell himself and others that most of the concerns expressed are mainly due to a lack of scientific knowledge and understanding. The very problem is that all these groups have good and even moral reasons to defend their various positions as top priority. None of them can be called an egoist in the proper sense. Even the profit-oriented industrialist can argue that his success will contribute considerably to the general welfare - not to speak of the good arguments the scientist can present. It is the delicate task of public authorities, and so of the Commission, to take generally acceptable course between all these conflicting requirements.

10.2 The scene of conflict

What is missing is a form of interaction between the quarters concerned which will lead to a balanced and uncontested co-existence of the various positions. Most of the groups involved would see each other either as competitors in the market of public favour or as threats to their own goals and ideals. The tendency towards thinking and acting in terms of antagonism will be the higher between groups the more they are professionally organised and claim to represent certain public interests. The conflict between these groups is sometimes even higher than the contrast between those whose interests are represented by the groups concerned. The increasing politicisation of public interest groups (IG) brings an additional element into the debate. IGs are a kind of interface between the public and those who act politically or economically in biotechnology. They contribute considerably to the formation of public opinion in a way similar to that of political parties in other fields, and in many of the political discussions on debatable matters in biotechnology public IGs are the very opponents (or partners) of political decision makers rather than the public itself. Interest groups can be considered as highly specialised political parties; and, like these, they would hardly retire from their business when their goals proclaimed have been achieved. Public interest groups vary considerably in character, ranging from a strong and fundamental opposition against nearly any research into gene-technology or its application, to the rather moderate and flexible position of many consumer groups.

The scene in biotechnology can be considered to be divided in mainly three parts: 1. Those who have a commercial interest in biotechnology and its applications (Industry, agriculture etc. and most of the R&D concerned), 2. those who deal with biotechnology for political and social reasons (comprising public authorities as well as public interest groups) and 3. the general public as consumer of the beneficial biotech goods and services as well as "consumer" of the risks and the more general socio-economic consequences involved.

The complexity of the biotech scene is based on the complex interplay between these groups. One possible interplay is the exchange of scientific, technological, economic and other relevant factual data, information and arguments with a view to the elimination of misunderstandings and the possible rationalisation of conflicts. It is widespread understanding rooted in old democratic traditions that this is the main, if not the only way to come to stable and reasonable forms of co-existence and compromises. Particularly the English culture thinks and acts in terms of a consensus which has to be found for all controversies and will be found if there is sufficient room for informed discussions. This is the very root of the idea of public information: the more people are informed the more successful will be their decisions - or as Mark Cantley (2) said: "If there is 'ignorant democracy', control without understanding, there is danger not only to science and technology, but ultimately to the society itself". A similar thought was expressed by Sir Walter Bodmer (3) in his famous UK Royal Society report 'The public understanding of science': "in the absence of widespread understanding we will shy at kittens, and cuddle tigers", i.e., we will be unable to manage benefits and risks of science appropriately. A third statement of that kind comes from Jon D. Miller (4), director of the US Public Opinion Laboratory: "Throughout the world, the importance of a scientifically literate workforce is recognised by political and economic leaders, and an increasing number of leaders in democratic societies have recognised the essential role of scientific literacy in the performance of citizenship responsibilities. Most governments of major industrial nations have strong commitments to improving or sustaining the quality of their programmes in science and mathematical education. Many nations are seeking to expand adult informal science education to maintain the levels of scientific literacy attained through the common schooling experiment". The view that people should be informed as much as possible about everything concerning themselves and the society they live in, is implicitly based on two ideas:

1. An improving level of the layman's scientific knowledge will improve the quality of his judgement on the political decisions to be taken in science and related political matters.

2. What we have (or what we therefore should have as many people say) is a participatory democracy rather than a representative one, i.e. a society where the citizen who is expected to be as emancipated as responsible will evaluate matters of public interest on the basis of his own knowledge and experience and then is involved, directly or indirectly in political decision or control processes, instead of leaving publicly important decisions to the legislative and executive bodies he has elected just for doing this. As to science: science itself has brought about the idea of its incorporation in general education, and is by this confronted now with the problem that the public more and more would claim participation in the definition of what research should be permitted and what forbidden.

Both ideas are more or less generally agreed. It is evident that people should take any opportunity to qualify democratic decision procedures by means of their own knowledge and that to improve this knowledge is their first and foremost task. But it has to be seen as well that factual scientific knowledge is just one of the factors determining people's attitude towards science. It might be plausible particularly for those who are used to think in scientific terms and who are proud of their scientifically trained intellectual self-control, that knowledge of science and attitudes towards science are positively correlated, i.e. that people would appreciate science and its applications the better, the more they know about it and the more they understand the mechanisms involved. This is rooted in the traditional idea that science per se is the most distinguished tool to achieve improved living conditions for all men. I do not contest that science indeed is the most powerful (and in many cases the only) instrument to solve certain human problems -particularly those the application of science has brought about itself. But this does not determine the view on the desirability of specific developments in biotechnology - neither with the public in general nor with scientists. Even fully expert academic biotechnologists who hardly suffer from a lack of knowledge can have diametrical views on the social risks and benefits of certain matters in their own field, as demonstrated impressively by the experts hired by the various groups. How, then, can we expect the view of even well-educated laymen to converge towards a reasonable and general consensus?

10.3 The social dimension

All this underlines that the existing conflicts between the producer and the consumer of risks and benefits of biotechnology can hardly be solved by just teaching the scientifically uninformed. It is not sufficient to tell people that biotechnology is probably the only instrument to fight successfully cancer, or to explain why the fear that biotechnological research may result in dangerous genetic monsters is unfounded, or why the deliberate release of genetically modified organisms is hardly the kind of threat to the environment that many people believe. Independent from whether this were correct or not - all this is not sufficient if the actual comprehensions, after all, are immunized against special scientific or otherwise factual information in the sense that people do not trust the information source concerned, i.e. if people do not believe in what is being told to them, or if the opposition against certain aspects of biotechnology is based upon culturally acquired ethical positions which are widely resistant against all non-ethical arguments. Particularly here it is evident that factual information would hardly dissolve objections, and that efforts to improve trust in the reliability of informational sources or regulatory measures would be of little help.

Let me explain the non-scientific character of the relation between science and the public in some more detail. It is comprised in the definition of any individuum that it has to cope with its environment. This applies to the most primitive animals paddling around in their pond as well as to men. In the beginning, the problems to be mastered were first of all physical problems: to identify and maintain food, energy or other life resources, to protect oneself against cold and other inconveniences of nature, to fight diseases and, where men are concerned, to improve the limited physical capabilities of our species by means of machines, computers and science at all. This became the very paradigm of occidental science: to understand nature in order to master it, where nature was understood as the physical environment of individuals.

But if we look around nowadays at the environment we have to cope with we will find out that the relevant aspect is shifting more and more from the physical to the social dimension. If we look at a usual day's course we will see that about just one or two percent would concern really physical problems and their solution. The major part will pertain tasks we have to accomplish in the context of a social rather than of a physical environment. Most of our daily efforts will not be honoured by nature, i.e. they would not help us to survive in deserts or rain forests. They rather have to be appreciated by our society which, in turn, will provide us with the goods and services we need for living.

Even if we deal as natural scientists explicitly with the problems given by nature, we usually do not do it in order to survive better in the physical world but to survive better in the academic quarters of our society (5). The selection forces we are subject to are social in character, not physical. The world we live in is first of all a social world. This applies even to those problems which are obviously physical in character such as the environment to be protected. What does that mean? In all nature one can find what could be called the phenomenon of risk homeostasis. Species, individuals or societies which have developed a new technique to solve a special problem in order to reduce the risks related to it, usually exploit the new possibility in a way that the total risk they are confronted with will rise again, after a certain time, to the previous level; the ruthless exploitation of strategic resources, so to say. A typical example is the car driver who uses the anti-lock brake not in order to reduce the risk of driving but to drive faster and more riskily. Insurance companies have reported that ABS drivers have sometimes an even higher accident rate than ordinary drivers. This, unfortunately, would hold even if an ingenious invention would allow us to cut in half the total environmental output of all production. After a while, I am afraid, we would take that opportunity to double our production. Another dreadful and very delicate example is the food and agricultural help for the most starving overpopulated regions in our world, if this aid will be used to produce new starvation in the form of new children. This, again, is a social and not a physical problem. The only real relief would be to break the circle of risk homeostasis, i.e. to redefine the priorities of our life strategies from short- to long-term aspects. This is why I called environmental risks a social problem. A longterm solution can be found only on the basis of social arrangements rather than by means of new technical development. Of course, this does not mean that the scientific environmental research such as presented here in such impressive quantity will lose its legitimation. Too many of today's environmental damages can be repaired only by means of special hightech measures. But we should take care that, in the long term, the environment will profit from it and not the satisfaction of other individual or social short-term demands.

If we go into schools and teach children in science and particularly in biotechnology we should tell them at the same time that science is not only the never-ending source of beneficial goods and services provided we succeed in managing the technical risks related to it. Science has to be seen in the greater context of, and in competition with, the other instruments we use to manoeuvre our society. This view must not be confused with critical positions on science based mainly upon the apprehension that there are physical and technical risks and dangers related to it and which we cannot keep under sufficient control, and that the best way to escape these risks is to refrain from the special research in question. We must not discuss here the actual risks concerning genetic monsters and the deliberate release of genetically modified organisms (GMO) and to what extent the arguments used are scientifically or otherwise reasonable. These are technical problems, as technical as the benefits are scientists and industrialists speak about. I believe that control, self-control and the many regulatory measures we have or we can develop are well suitable to a successful risk management. So the balanced account of science is or will be by far positive. The danger I see in science is that it may monopolise our thinking in the sense that we consider science as a more or less omnipotent tool which would relieve us from the need to reflect on other tools. Our social responsibility does not end at providing society with a well-running science. We rather have to define the reference system of values according to which we will respond to the possibilities of science. Or in other words: We have to think in long-term categories in order to escape the circle of risk homoeostasis, and this is more than just organising the development of scientific solutions for technical problems.

The widespread (though now diminishing) belief in the overwhelmingly positive potential of science is rooted in our general belief in the power of rational thinking. I am not going to say that there is any reason to resign rational approaches in problem solving; but we have to be aware, that the high prestige of rationality is mainly due to co-evolution of rational capabilities and their applications, i.e. due to the fact that we favour just those goods and values which can be realised only by rational and scientific efforts, which, in turn, will increase our dependency on the further development of these capabilities. Cultures where the achievement of a good relationship with God ranks above the acquisition of material goods and technological achievements, may less depend on the extension of rational skills. It is a general phenomenon in both organic and cultural evolution that capabilities and skills (and therefore organs implementing these capabilities) can be evaluated only in the context of a certain application (6). Whether a small or a big bill is better for a bird cannot be said without knowing what is the bill for: picking grains, cracking nuts, climbing trees or fighting. Nor do rational or even scientific capabilities represent intrinsic merit. They can be weighed only with respect to their (potential, intended or actual) application. Particularly it cannot be said that species with rational competences will represent an a-priori higher fitness than others. In view of the high number of crucial human problems based on a lack of social coherence, it may well be possible that societies where unconscious and therefore irrational problem solving capabilities would dominate, will master their future better than we are able to do.

10.4 Knowledge and attitudes

The need to think in longterm categories, comprising both scientific and social aspects, is also the reason why the CEC calls its efforts towards a better relation between the various quarters involved in biotechnology, the "Socio-economic Integration of Biotechnology" rather than just "Public Information". This reflects the view that the ongoing conflict between research, industry and wider parts of the public on particular biotechnological issues cannot be reduced to a kind of misunderstanding of science which could be healed just by more and better information - as, unfortunately, too many people still believe.

One tool to proceed in this matter is a extended communication between all involved. "Extended" means that not only scientific and technical data are exchanged but also data on the economic implications, the legal and regulatory background and on the social and safety aspects. This requires us to provide platforms for dialogues in their various forms. The Commission has held several workshops with experts concerned and in collaboration with consumer organisations from Europe as well as from the

U.S. But it requires also the elaboration and evaluation of the methods to be applied, based on our own research and analyses.

For this the Commission feels the need to have more detailed knowledge on better methods and strategies for improving the relation between the various quarters in the biotech scene. To deal with methodological questions and to try to improve the methods concerned is not only a matter of more or less effectiveness of public information. Methodological considerations can be of high qualitative importance in sofar they can inform on whether a special measure is likely to be productive or counterproductive. This, for example, concerns the relation between knowledge and attitudes. Is it true, as many people believe - particularly from the side of the natural sciences - that scientific knowledge will determine more or less the attitude towards the science in question? If this were true, we could, of course, confine ourselves to public information in the usual sense and through the usual channels such as the media. But as we can learn from the social psychologists, in some cases the relation between knowledge and attitude can be just the reverse. Then attitudes are the primary variables, which will select the eventually circulating information.

It is wide-spread understanding that the key notion to describe the relation between science and the public is the knowledge about benefits and risks related to science and its applications. Identifying benefits and risks of biotechnology objectively and informing the public accordingly is expected to minimise public concerns and objections. Controversial opinions which scientists would call "irrational" are assumed to be mostly due to the lack of appropriate factual information on both the scientific and the legal (regulatory) aspects which will eventually and in the long run determine public attitudes. The relation between information and attitudes is assumed to be that of cause and effect. This is the basis of many, if not of most measures and campaigns to improve the relation between biotechnology and the public. The term "Public Information" (PI) which has come to stay as the general label for all these activities would suggest by itself the causal link between information and attitude.

This approach, however, neglects that the effect of messages, data or other kind of information depends on both their content and their interpretation by those perceiving them rather than on the content alone. Even more, the relation between information and attitude can be just opposite: People will select or reïnterpret the information available according to their existing attitudes so that eventually only those pieces of information will be publicly discussed (and therefore disseminated) which fit into (or reinforce) current attitudes and prejudices.

This is underlined by what Dorothy Nelkin has written (7): "Many scientists still believe that the media are responsible for negative public attitudes towards science, that the tension between science and society reflects the poor public understanding of science, and that an adequately informed public would share the enthusiasm of scientists themselves. Thus, they try through public relations to convince journalists to project a more favourable public image. But this belief oversimplifies the complexities of public attitudes towards science, and underestimates the importance of pre-existing attitudes in shaping readers' interpretation of media images". This is also confirmed by Brian Wynne, University of Lancaster, when writing(8): ". . . But biotechnological understanding has been conventionally seen as a natural good - like any other form of knowledge. Our research has shown that lay people respond to scientific information not at all in a purely intellectual way. That is, even people capable of assimilating an offered level of technical knowledge may resist it, because they sense in that knowledge, not a morally or socially neutral and detached 'free good', but a trojan horse of associations, with technological, social or moral visions and future trajectories that cause anxieties. Of course the associations may also be positive. But the frequent lack of articulation of these 'deep structures' underlying 'neutral' packages of knowledge confuses and perhaps exacerbates negative public reactions. Thus even liberal information programmes may backfire if these tacit dimensions are foreclosed".

In order to put these and similar views on a more systematic basis we organised a workshop on "Knowledge, Attitudes and Behaviour towards Biotechnology" (Brussels, Sept. 10, 1990) with experts from the various fields involved.

The workshop came to the following conclusions:

1. One of the basic problems is the lack of trust in science, scientists and application by industry rather than the lack of scientific or technological knowledge. It is important to tell people that science is not something which presents ready-made solutions but that scientists have similar problems and concerns to those of the man in the street. What confuses people is the claim of absolute competence of science. What is needed, therefore, is communication rather than information. People have to be convinced

that the biotech actors not only act and speak but also listen. Information should not only concern science and technology and their applications but also economic, legal and regulatory aspects to increase trust.

2. Communication activities have to consider that the effect of information is strongly source dependent. Dissemination of knowledge, therefore, has to apply appropriate mediators (non-governmental and non-industrial organisations such as universities and museums). Appropriate funding procedures for independent communicators have to be identified which will not lead to a loss of public credibility.

3. Scientists who have sometimes a rather vague idea of what the effect of their communication could be, have to be trained in communication sciences and they have to be aware that public communication is part of their success as much as finding public money is. Scientific communication activities aiming at the public have to be professionalised in the sense of PR, in co-operation with social scientists and psychologists (anecdotal approach, approach via actual problems to be solved by biotechnology). A particular point is targeting and segmenting: different strategies have to be developed according to the various target groups as well as to the various subjects in question.

4. Concerning the high importance of cultural differences, the state of the art in public knowledge and attitudes in the various countries have to be analysed. It was stated that the role of consensus and the awareness of need to find it, is rather different in the various Community countries. This may be one of the reasons why public debates in public groups on biotechnology and its implications generally came to peaceful conclusions in England whereas they did not lead to comparable results in Germany.

So we know that knowledge and attitudes are not necessarily positively correlated. But we do not know in what case there is a negative correlation or no correlation at all. This question is crucial as we mentioned above the assumption that the effect of information mainly depends on the interpreting pre-existing attitudes. If this is correct, we cannot expect any correlation - except the case that the attitudes expressed themselves are the result of previous scientific information. For the politician or the industrial PR manager it does not make a difference whether a negative attitude towards biotechnology is due to a previously perceived information which was scientifically wrong and insufficient or due to experiences concerning trust, credibility and related topics. What is important is only whether existing attitudes can be influenced better by scientific or otherwise factual information or better by contextual measures dealing with trust etc.

In order to get the empirical evidence needed the Commission organised April 1991 an opinion survey on "Public Perception of Biotechnology" on the basic of 12.500 interviews in the Member States of the Community (9). The result confirms that there is an additional positive correlation between knowledge and attitude towards biotechnology if other parameters indicate positive attitudes as well. Further to this, however, there seems to be a complex and culture depending cross-correlation between the various variables. In Denmark, for example, people regard biotechnology as a highly risky matter. Nevertheless, the support of further research is recommended. This is coherent only if people would trust in the public authorities controlling risks and potential misuse of biotechnology. This, indeed, is confirmed by the survey. There is no other country where people believe so much in the reliability of information released by public authorities. The German population as well is afraid of biotechnological risks, and as well trusts public information sources far above the European mean value - but is strongly against further backing of research in the field. These and other open points are subject of ongoing Commission analyses particularly on cultural differences .

I started with the metaphor of Beauty and the Beast in order to characterise the various conflicts we are confronted with. I hope enough evidence was presented that there is neither a beauty nor a beast, i.e. that the conflict is not of the kind that each of us can unambiguously say how beautiful or beastly he is. Science, as Niklas Luhmann has shown, and particularly biotechnology, is not a subsystem of our society in the sense that it could be regarded separately from other subsystems. This insight, of course, is not a surprise and it may not provide us with any particular means to improve the relations between the various quarters of the biotech scene. But understanding that this is as it is may help us to identify and eliminate approaches which, at least in some cases, are counterproductive rather than helpful.

References

1. Press release 18. April 1991, by SAGB (Senior Advisory Group on Biotechnology), CEFIC (European Chemical Industry Federation)

2. Mark F. Cantley, "Democracy and Biotechnology", *Swiss Biotech* 5 (1987) Nr. 5, p. 5-15

3. Royal Society Working Group, under the chairmanship of Bodmer, Walter. *The public understanding of science,* The Royal Society, London, 1985

4. Jon D. Miller, "Empirical Comparisons of Public Understanding of Science in Japan and the United States", Public Opinion Laboratory, presented to the 1991 annual meeting of the American Association for the Advancement of Science, Washington, D.C.

5. O. Diettrich, *Kognitive, organische und Gesellschaftliche Evolution*, Parey, Berlin und Hamburg, 1989

6. O. Diettrich, "Realität, Anpassung und Evolution", *Philosophia Naturalis*, 1991, II, p. 147

7. Dorothy Nelkin, "Science in the public eye", a review of "Making Science Our Own" by Marcel C. LaFolette, *Nature*, Volume 348, p. 121, 8 November 1990

8. Brian Wynne, letter to the CEC, 1989

9. EUROBAROMETER 35.1 "Biotechnology" for the Commission of the European Communities, DG XII, "CUBE"-Biotechnology Unit, by INRA (EUROPE), European Coordination Office SA/NV, June 1991

11 SCIENTIFIC CONTROVERSIES IN FOOD BIOTECHNOLOGY

Piet Schenkelaars
Friends of the Earth, Brussels, Belgium; Contactgroup Biotechnology, Wageningen, The Netherlands

11.1 Introduction

Our 'Contactgroup Biotechnology', a non governmental organization, supplies information on biotechnology and world foodproduction to other mainly Dutch non governmental organizations, such as farmers- and rural women organizations, environmental-, consumer-, third world- and animal welfare groups. My two colleagues finished studies in agronomics and sociology. Speaking for myself, I got a scientific training in molecular biology and in philosophy of science. In our group we try to analyze what the impact of biotechnology in world food production will be on the quality of the food, the environment and the labour in agriculture and food processing industries, on the possibilities for self-provision of third world countries and on the distribution of the societal income, earned by implementing this biotechnology. Our goal is to inform public groups about what the genetic engineers are doing in their laboratories, what kind of market strategies food industries are following and what kind of policies, laws and regulations are being formulated by inter- and national bureaucracies. Information about these issues -we hope- should enable these groups in taking position in the discussions on several aspects of biotechnology.

At the moment we participate in three campaigns against the introduction of Bovine Growth Hormone, against deliberate releases of genetically manipulated organisms in the environment and against the patenting of genetically altered plants and animals.In this vast field of biotechnology and food production we encounter lots of scientific controversies and questions how to regulate properly this 'risky genetic engineering business'.

In this contribution I will mainly focus on such controversies about the impact of biotechnology on the environment and on food quality. My approach is rather journalistic and is based on interviews we held with several scientific experts.

11.2 Repair of the environment

Advocates of biotechnology promise to solve the ecological problems caused by current agricultural and industrial practices. Transgenic crops with built-in resistances against diseases and pests and genetic manipulated bacteria or viruses should replace chemical pesticides. In processing industries chemical reactions, which often require high temperatures and pressures, should be substituted by microbial or enzymatic pathways, which consume less energy. Cleaning up polluting waste streams from agriculture and industry, eventually using them as substrate for products, are other recommended possibilities.

Critics agree that current ways of producing food cause damage to the environment, but they point out, that almost nothing is known about the ecological risks of genetically manipulated organisms. As alternative they often advocate integrated pest management and organic farming as safer options for solving these environmental problems.
But these options are of less interest to the 'genetic engineering business'. And the argument that a new class of risks is being introduced into the environment and society, gets twisted around and becomes a starting point for the development of risk- molecular biology and technology assessment procedures, which uptill now implicitly exclude research programmes on alternative solutions.

The scientific-technical debate on the ecological risks of genetically altered organisms shows some serious problems in the communication between the several disciplines in biology, which are needed to make scientifically thorough risks evaluations. These problems manifest themselves on the cognitive, as well as on the social or institutional level.

The discussion about the risks of using recombinant-DNA techniques more or less started with the 'Berg-letter' in 1973. In the years thereafter several conferences took place. The Asilomar Conference got much press coverage and public attention, because the genetic engineers imposed a unique moratorium on themselves, which lasted for about twelve months and which people as James Watson later regretted.
In his book "The double edged Helix" the biochemist Liebe Cavalieri describes this decision of the scientific community as a defense mechanism against interference by politicians from the American Congress. The genetic engineers had become concerned that their endeavourers were going to be constrained and this became very worrying in an international

competitive environment. In those twelve months the genetic engineers quickly developed a consensus on precautions measurements and safety rules, which reassured the politicians and the public. This consensus, some argue, was a false consensus, not really based upon much thorough scientific investigations.

The ecologist Philip Regal has made some interesting observations at this Asilomar Conference:

"There were many concerns raised about social issues, about economic issues, about international relationships, about the misuse of nature, about the effects on university and on the honesty of science, but one issue was raised, that they fell, should focus on. This was the issue of biohazard; the escape of an organism from the laboratory, which then would cause a worldwide disaster. That was rather phoney, because most laboratory organisms are like white mice, so highly inbred, that their chances of living outside the laboratory are very slim. No ecologists or evolutionary biologists were involved at that time. If you are cynical, you could say, the reason they phrased it so unrealistically was, that people did not care to get involved. If I was interested and stepping in and saying: I am concerned about an intromitted strain taking over the world, I as a scientist would have looked very foolish. And if I was worried about a small laboratory accident, then they would say: Well, you are making a value judgement. And then I would like I was being political. In terms of your professional stand as a scientist, you are tempted not to get involved in it. So, most ecologists said, the molecular biologists got themselves into it, so let them deal with it themselves. And one thing you better realize is: The molecular biologists came into biology from chemistry, they are not in the same network of people as the ecologists are, they are like a foreign colonizing power."

But that situation changed about 1984. By that time it looked the genetic engineers were no longer dealing with white rats, i.e. with organisms, which are genetically mutilated by so many years of inbreeding. Now they were able to create organisms, that could live in nature on their own and that is of concern to ecologists. Philip Regal:

"I was particularly concerned, because I was on a committee at the university, that had to review new degree programs, one them was in genetic engineering. They said it was going to be the best programm in the world. I asked them, how can it be the best program, if you are training people to know only chemistry, they won't know any ecology, any evol-

ution, any ethics, any economics. I said where is society to get advice on how to use these new forms of life, if the people who are creating them, don't know anything besides biochemistry, no social sciences, no ecology, no politics. Well, they said, our job is only to build them.
So I became very concerned. I made phone calls and this was true all over the world. They were training people, who only would know how to modify living forms and put them out in nature, but they would not know what these organisms would do, once they released them, what the effects on the environment or on human health would be. The training was impossibly narrow and it was not going to qualify them to make competent judgements.

So I started to investigate why they thought it was so safe to be ignorant. And I found out they used a series of arguments to argue that any genetically engineered organism ought to be safe. But they were based on 19th century biology, on ideas that we had thrown out of biology decades ago."

In 1984 and 1985 several joined conferences of molecular biologists and ecologists were organized, because within the next decade organisms would be created, that should live in nature. The conclusion was that there were things to be worried, but the ecologists could not say exactly what the problems were and how to deal with them. They only could say: One has to be very careful.

As a result of the ongoing discussion, a quite important article by the Ecological Society on "The planned introductions of genetically engineered organisms: ecological considerations and recommendations" appeared in 1989. It is remarkable that the conclusions of this article do not strictly follow from its analysis. According to Sheldon Krimsky of Tufts University Boston, due to an underlayer of other issues going on besides:

"There is the question of the ecologists trying to assert some control over this area of knowledge. There is the issue of ecologists trying to maintain a certain respectability in the sciences amidst the geneticists, who have much of the funding and who get all of the Nobel Prizes. So you got this playing out in a sense. Here you have issues that are overlapping disciplines and people in the different disciplines are asserting competence to make judgements in this area. So the ecologists are trying to find some common grounds among their group, but they do not want to seem too radical. They want to set themselves up in such a way that they can be critical, but not too far out, because then they would be dismissed completely by other elements of the scientific community."

But there are also other elements in the underlayer of this controversy about the ecological risks of genetically engineered organisms. Tremendous efforts are made to link genetic engineering up to industry. It means the scientific agenda becomes constraint. Commenting on this Peter Wheale and Ruth McNally of Bio-Information London state that:

> "Certain questions won't get bothered to ask. These questions would have concerned perhaps the ecologists. Off course, under the regulations something has to be said about this, but compared to serious, well funded, extensive and curiosity-orientated research on eco-systems, there are no accolades for extensive risk-assessment. Very little money and very little applause for that kind of research. We also think that the genetic engineers of the eighties have learned genetics in such a static and reductionistic way; they are probably even shielded from the fact there was a scientific controversy in the seventies."

The first deliberate release experiment in the USA took place with the so called Ice Minus Pseudomonas. In his study on this case Sheldon Krimsky could find no ecologists who specifically say that the Ice Minus case was potentially hazardous; the problem was the things that are coming after:

> "My concern was that the Environmental Protection Agency would be particularly careful with a case that most people agree was not the most dangerous case. What happens down the line, when the cases get more risky and the immediate public attention is not directed at that, then EPA will fall back to a much more relaxed position. EPA is bound by an ideology that you should not be too harsh on business."

Field release experiments are being conducted in order to investigate the ecological risks of genetically engineered organisms. In the UK the first field release with genetically altered Baculovirus was done by Steven Bishop from Oxford University in 1986. The objective of the research programm was to assess the consequences of their deliberate release into the environment, although in the future this application of gene-technology should lead to the development viral pesticides to replace chemical pesticides in agriculture. There is no doubt about Bishop's integrity, but according to Ruth McNally:

> "It was not particularly realistic for other experiments that might follow. What it was showing was a model experiment. You cannot critise someone for that. Except you find it hard to relate that to a real pest in the real world. The experiment was done in the autumn, when there would be none of the caterpillar-species around, so the virus could be controlled, because this virus needs that organism to reproduce. It was a safety feature, but it also invalidated the experiment as a preparation for a real test of a real pest. So, one reaction could be: The experiments were good, but they have nothing to do with real application of genetic engineering in the environment. The other criticism would be: Okay, they have done a set of experiments which are supposed to prove that genetic engineering is very safe. One of the things you see in his paper is that a large amount of viruses disappeared from the plot and he does not have a good explanation why that happened. It is interesting that he is now expecting money from the European Commission to do risk assessment work. This is always the EC-response to criticism: We've got risk assessment money. But we think that those risk assessment programmes are designed to show there is no problem, rather then to find out what the problems are."

The technical discussions about the risks resulted in "Recombinant DNA safety considerations" by the OECD in 1986. These considerations functioned as a basis for the formulation of regulations worldwide. Last year the EC adopted two guidelines on the contained use of genetically modified organisms and on the deliberate release of genetically modified organisms into the environment.

In the Netherlands there now also exists a legislative framework to regulate the use of genetic engineering. Within this framework the 'Commission on Genetic Modification' (VCOGEM) plays a crucial role; this Commission is a technical body which advices the Minister to authorise or not to authorise an scientific experiment or commercial application with recombined genetic material. Beforehand a risk analysis has to be made. One of the questions for this analysis to be answered is: What are the chances that the genetic engineered organism or its offspring causes harm to the environment? The real problem with this question is the meaning of the notion 'harm to the environment' in it, because this notion cannot be filled in by scientists only.

Last year one of the members of this Dutch committee wrote in a report on "The risks of transgenic plants for wild plants":

"The acceptability of a risk can only be established in political weighing procedures."

To my opinion this is a quite amazing statement from a scientist to acknowledge the political content of his work. Nevertheless, the members of this committee are appointed because of their scientific expertise; nobody represents a democratic political party or organization. And at the moment there are no signs that this situation will be changed. So, the decision-making in these authorization procedure remains rather technocratic.

Besides, this committee is dominated by biologists from molecular disciplines, who assess the risks without having any ecological models with predictive power, and without sufficient ecologists and evolutionary biologists as members.

11.3 Food quality

In an international competitive environment industries view technology as a risk-reducing activity. Bio- and genetechnology have interesting opportunities to offer to food industries. Because the price of raw materials forms a substantial part of the costprice of an endproduct, raising the productivity of agriculture by manipulating crops and cattle may be profitable. Any decline in the use of agricultural raw materials and/or any increase in the diversification of commodities used in processing means a drop in farm prices and a loss of farmers. These savings at the processing level may or may not be passed along to consumers. And rather than eliminate additives, the new techniques may replace chemicals with new life forms causing even more regulatory complications than the old additives.

The patenting of the technology will lead to yet greater market concentrations and oligopolistic pricing. Increasingly, farmers will be sold patented breeding stock, plants and animals, by food processors. The same food processors will buy the harvest. In many cases, the entire relationship for both the input and the output will be contractual.

The ideal for processing industries is to refine the agricultural raw materials into carbohydrates, proteins and fats. And hereafter these molecular components get reassembled with means of additives into an endproduct further on the production line. Advocates of biotechnology

promise to enhance the quality of foodproducts, for instance by replacing chemical processing steps by enzymatic or biochemical reactions. Chemical additives will be substituted by biotechnologically produced flavours, conservants and thickeners, suggesting these are 'natural' additives.
Food processors mean they need genetic engineering to produce new flavours, because the taste-intensity of current foodproducts got lost by product-rationalization and harvest-technologies.

The science of human nutrition also thinks in molecular terms. Humans do not need food, they need the right combination of carbohydrates, proteins, fats and minor nutrients. This way of thinking about human nutrition has been developed over the last century and at the moment even the population thinks in the same terms; for example, an orange becomes reduced to a dosis of vitamin C.

This approach of human nutrition is a prerequisite for implementing genetic engineering techniques in foodproduction. In this approach it is possible to define genes, which produce the required molecular components and to manipulate those genes for the production of these components in transgenic crops and animals.

Unilever, which is the world's largest processor of edible fats and oils, invests lots of money in raising the productivity of oil-containing crops, and in the manipulation of the composition of these oils, for instance, a higher content of short chain of poly unsaturated fatty acids in sunfloweroil or the elimination of euridic acid and glucosinates in rapeseedoil, but also the interchangeability of the oils and fats in endproducts is constantly being investigated, especially enzyme engineering offers new possibilities to processing industries.

In the past these strategies enabled Unilever to dominate the margarine-butter fight against the dairy industry. At the moment the new technologies are used by Unilever to produce for example to make a cheese-like product, in which the milkfat has been replaced by soybeanoil. In this way Unilever challenges the dairy industry again, and so the dairy industry feels itself forced to develop this technology in order to keep or enlarge its marketshare. For the dairy industry this is rather problematic, because most of these industries still are based on the cooperative membership of dairy farmers. In this setting it is not helpful to their members to use raw materials coming from elsewhere.

This product may not be sold as cheese, but looking at the advertisements it is being presented as cheese. And because lots of consumers nowadays suppose that vegetable oils are better than animal fats this product gets an aura of being healthier than normal cheese.

It is interesting to look at some advertisement slogans Unilever used in selling margarine with a high content of poly unsaturated fats. About ten year ago these margarine were 'good to heart and blood vessels'. A few years later it became forbidden to use this slogan, because there were no hard scientific data to conform this. So the slogan had to be changed and at the moment this Unilever product 'helps to lower the level of cholesterol'.

At the moment human nutritionist are investigating questions like: Which poly unsaturated fats are the best, those in cis- or those in trans-formation?

Another argument for using bio- and genetechnology in foodprocessing is the diversification of consumer goods. But this is a diversification of processing processes. The genetic diversity of plants, animals and microbes used in foodproduction becomes more and more constrained, because of the processing demands on the composition of the agricultural raw materials. On the one hand, this process of genetic erosion threatens future plant- and animal breeding, and on the other hand the content of the food-products becomes more and more standardized.

According to Horst Grimme of the University of Bremen this kind of food does not stimulate the immune-competence of the population, due to the lack of biological diversity and natural complexicity.

The use of chemical additives in foodproduction has made consumers suspicious, although the use of these compounds became more strictly regulated. Procedures for assessing the risks of the separate compounds have been developed. Safety experiments are performed by testing the additives on animals, especially on rodents. But there always remains the question of the applicability of these data to humans. Besides, the additives are tested separately, and synergistic effects between the many compounds a person consume in a day are almost impossible to investigate.

Compounds can be also tested on bacteria for their toxicity and their ability to damage DNA. Recently Bruce Ames, who invented this test, declared to doubt the validity of his own method and of the animal experiments, thereby questioning the whole regulation-system. But at the moment nobody knows alternative ways of evaluating the human health risks of additives.

The safety evaluation of the use of Bovine Growth Hormone in dairy production shows some serious flaws. According to Samuel Epstein data on the safety of the milk and on animal welfare have been manipulated by the BGH-producers like Monsanto, Eli Lilly, Upjohn and Dow Chemicals:

"These are self-interested industries with little or no social conscience. Their pre-occupation is exclusively on short term economic interests. So, the data-base on BGH has been entirely generated by the industry, either by in-house scientists or by university dairy science departments under contract with these companies. There has been no independent research. And the Food and Drug Administration over the last thirty years has been consistently more interested in having a close relationship with the food and chemical industry and not in protecting the consumers. Six years ago before most of these data on milk safety and on animal welfare were even available, the FDA allowed the sale of this milk from clinical trials without any labelling, because they said it is safe, even in the absence of any data."

11.4 Conclusions

I have tried to sketch some scientific controversies in the field of biotechnology and foodproduction. These controversies also contain economical, political and ideological elements.

In connection to this I would like to end with an observation made by Ruth Mcnally and Peter Wheale:

"You can stand up and say all exaggerations you like about the benefits of science and nobody will mind. But if you suggest there are dangerous sides, people get very upset and start calling you a scaremonger. When you are talking the benefits, you can take minute examples and on the strength of one example, you can promise the whole of genetic engineering. When you talk about the dangers and you take a specific example, then that is just one thing. You are not allowed to generalise.

Sometimes they use very broad concepts, like 'all experiments carry risks, therefore the fact that you discuss risks means nothing: Do you want to stop science? As soon as you start to say No! to certain possibilities, you are immediately anti-science and irrational. It is a very clever discussion technique!"

INDEX

THEORY AND DECISION LIBRARY

SERIES A: PHILOSOPHY AND METHODOLOGY OF THE SOCIAL SCIENCES

Editors: W. Leinfellner (*Vienna*) and G. Eberlein (*Munich*)

1. G. Zecha and P. Weingartner (eds.): *Conscience: An Interdisciplinary View*. Salzburg Colloquium on Ethics in the Sciences and Humanities (1984). 1987 ISBN 90-277-2452-0
2. R.W. Scholz: *Cognitive Strategies in Stochastic Thinking*. 1987 ISBN 90-277-2454-7
3. H. Nurmi: *Comparing Voting Systems*. 1987 ISBN 90-277-2600-0
4. M. Schmid and F.M. Wuketits (eds.): *Evolutionary Theory in Social Science*. 1987 ISBN 90-277-2612-4
5. C.L. Sheng: *A New Approach to Utilitarianism*. A Unified Utilitarian Theory and Its Application to Distributive Justice. 1991 ISBN 0-7923-1301-1
6. F. Forman: *The Metaphysics of Liberty*. 1989 ISBN 0-7923-0080-7
7. G. Bernard: *Principia Economica*. 1989 ISBN 0-7923-0186-2
8. J.W. Sutherland: *Towards a Strategic Management and Decision Technology*. Modern Approaches to Organizational Planning and Positioning. 1989 ISBN 0-7923-0245-1
9. C. Vlek and G. Cvetkovich (eds.): *Social Decision Methodology for Technological Projects*. 1989 ISBN 0-7923-0371-7
10. P. Hoyningen-Huene and F.M. Wuketits (eds.): *Reductionism and Systems Theory in the Life Sciences*. Some Problems and Perspectives. 1989 ISBN 0-7923-0375-X
11. K.G. Grunert and F. Ölander (eds.): *Understanding Economic Behaviour*. 1989 ISBN 0-7923-0482-9
12. G. Antonides: *The Lifetime of a Durable Good*. An Economic Psychological Approach. 1990 ISBN 0-7923-0574-4
13. G.M. von Furstenberg (ed.): *Acting under Uncertainty*. Multidisciplinary Conceptions. 1990 ISBN 0-7923-9063-6
14. R.L. Dukes: *Worlds Apart*. Collective Action in Simulated Agrarian and Industrial Societies. 1990 ISBN 0-7923-0620-1
15. S.O. Funtowicz and J.R. Ravetz: *Uncertainty and Quality in Science for Policy*. 1990 ISBN 0-7923-0799-2
16. J. Götschl (ed.): *Erwin Schrödinger's World View*. The Dynamics of Knowledge and Reality. 1992 ISBN 0-7923-1694-0
17. R. von Schomberg (ed.): *Science, Politics and Morality*. Scientific Uncertainty and Decision Making. 1993 ISBN 0-7923-1997-4

KLUWER ACADEMIC PUBLISHERS – DORDRECHT / BOSTON / LONDON

MIX
Papier aus verantwortungsvollen Quellen
Paper from responsible sources
FSC® C105338

If you have any concerns about our products,
you can contact us on
ProductSafety@springernature.com

In case Publisher is established outside the EU,
the EU authorized representative is:
Springer Nature Customer Service Center GmbH
Europaplatz 3, 69115 Heidelberg, Germany

Printed by Libri Plureos GmbH
in Hamburg, Germany